Sickening

SICKENING

How Big Pharma Broke

American Health Care

and How We Can Repair It

JOHN ABRAMSON

MARINER BOOKS

Boston   New York

marinerbooks.com

Library of Congress Cataloging-in-Publication Data has been applied for.
ISBN 978-1-328-95781-8 (hardcover)
ISBN 978-1-328-95698-9 (e-book)
ISBN 978-0-358-57836-9 (audiobook)

Book design by Kelly Dubeau Smydra

22 23 24 25 26 LSB 6 5 4 3 2

THE RESEARCH AND IDEAS PRESENTED IN THIS BOOK ARE NOT INTENDED TO BE A SUBSTITUTE FOR CONSULTATION WITH YOUR DOCTOR OR OTHER HEALTH-CARE PRACTITIONER. BEFORE STARTING OR CHANGING ANY MEDICAL TREATMENT OR OTHER REGIMEN, DISCUSS THE ISSUE(S) WITH YOUR HEALTH-CARE PROFESSIONAL.

To Charlotte, without whose love, friendship, indefatigable support, and fearless criticism this book would not be possible

Be not intimidated, therefore, by any terrors, from publishing with the utmost freedom, whatever can be warranted by the laws of your country; nor suffer yourselves to be wheedled out of your liberty by any pretences of politeness, delicacy, or decency. These, as they are often used, are but three different names for hypocrisy, chicanery, and cowardice.

— JOHN ADAMS,
A Dissertation on the Canon and Feudal Law, 1765

Modern medicine has been a powerful force for good, and many people owe their lives to that power. However, because of humanity's shared reverence for that success, combined with the increasing financial rewards from the industrialisation of healthcare, almost everyone has been slow to recognise that medicine also has great power to harm.

— IONA HEATH,
British Medical Journal, February 2020

CONTENTS

INTRODUCTION

Tragically, during the first year of the coronavirus pandemic, an average of fourteen hundred Americans lost their lives to COVID-19 every day. Far more tragically, but with far less public awareness, Americans have been dying unnecessarily at almost the same rate for two decades.* This invisible tragedy is occurring not because of a once-in-a-century pandemic, but rather because, compared to citizens of other wealthy nations, we in the United States have such inferior health and health care.

Not only is our health worse, but the pace at which we are falling behind is accelerating. For example, the (pre-pandemic) rate of American deaths that could have been prevented by adequate medical care was by far the highest: half again higher than the average of nine other wealthy countries. Similarly, the United States ranks lowest in quality and access to health care among eleven wealthy nations, and it is the only country to have declined on this measure since 2010, despite the expanded access provided by Obamacare. And since 2000, Americans' healthy life expectancy has plummeted from thirty-eighth in the world to sixty-eighth in 2019 (now behind China, Cuba, and Jamaica). Citizens of Japan live 8 years longer in good health and Canadians live 5.2 years longer in good health than Americans do. How can this possibly be, when Americans clearly

---

* In 2017, excess deaths in the United States equaled 1,356 per day, based on the difference between the age-adjusted U.S. mortality rate and the average of comparable countries.

have the best access in the world to the latest medical advances? The unvarnished truth is that the siren call of breakthrough medical innovation commands far more of our attention than the alarms set off by the decline in our health.

Making matters even worse, while we are spending 17.7 percent of our GDP on health care annually, eleven other wealthy countries are spending an average of just 10.7 percent. This additional 7 percent of GDP translates into our spending an *excess $1.5 trillion* on health care every year. To put this in more personal terms, despite our health losing so much ground in comparison to that of other wealthy and not-so-wealthy nations, Americans spend $4,500 *extra* per person each year — like an unlegislated tax — on health care.

Obviously, there is no single source of the dysfunction in our health-care system, but the most powerful force driving this toxic combination of poor health and high costs is Big Pharma's influence on American medical care. And the reason the pharmaceutical industry has been able to achieve this "tail wags dog" position is that, over the past forty years, public funding for clinical research and federal support of university-based medical research has declined, allowing the drug companies to step in to fill the gap. This has led to increased commercial influence over much of the information that doctors rely on to determine optimal treatment for their patients.

The pharmaceutical companies now control most of the medical research agenda, and their primary goal is not to improve Americans' health but to maximize their own profits, which they do masterfully. To that end they control the design, conduct, and analysis of most clinical research; and they largely control the delivery of the results of that research across the entire spectrum, from the most respected peer-reviewed medical journals to all those annoying drug ads on TV. Although the supposed purpose of this information is to educate doctors and the public, it is — truth be told — carefully curated to disseminate key marketing messages designed to maximize drug sales.

Still, how does it happen that so many smart, well-trained, hardworking, and dedicated physicians are misdirected by the commercially motivated "knowledge" produced by this self-serving system? Ironically, doctors are vulnerable to this misinformation precisely because they are taught to base their practice on the best scientific evidence published in peer-reviewed medical journals, respected clinical practice guidelines, and recommendations made by recognized medical authorities. But these trusted sources have become increasingly dependent on drug-company funding.

One of the best-kept secrets in all of health care — understood by few doctors — is that the peer reviewers, medical journal editors, and guideline writers, who are assumed to be performing due diligence to ensure the accuracy and completeness of the data reported from company-sponsored studies, *do not have access to the real data from these trials.* The published reports that doctors accept as fully vetted scientific evidence can be more accurately described as unverified data summaries prepared largely by or for the sponsoring drug companies.

For sure, some newly approved drugs — one out of eight — provide heretofore unavailable medical benefits. These can be genuinely lifesaving or quality-of-life-improving, like the drugs that transformed HIV/AIDS from a death sentence into a chronic disease compatible with a normal life, drugs to treat hepatitis C, and drugs to treat (but not cure) cystic fibrosis. But unlike other wealthy countries, the United States lets drug companies charge as much as they want, so the drugs that offer unique benefits are generally priced at ransomlike levels. Moreover, because the industry controls much of the scientific evidence that reaches health-care professionals and the public, the seven out of eight newly approved drugs that do *not* provide previously unavailable benefits can be promoted as if they do. The business environment for prescription drugs in the United States is so different from that of other wealthy countries that an estimated two-thirds to three-quarters of global pharmaceutical profits come from the United States.

Americans are well aware that Big Pharma is taking advantage of them. A Gallup poll conducted six months before the COVID-19 pandemic began found drug companies to be the least well regarded among the twenty-five industries included in the survey, Pharma's worst ranking since the annual survey began, in 2001. This strong negative sentiment reflected rapidly increasing drug prices and occasional scandals, most recently the gross overselling of prescription opioids, which has contributed to tens of thousands of American deaths each year.

Big Pharma does not set out to purposely harm Americans' health, but its primary job has become the exploitation of each situation as a unique opportunity to maximize profits, regardless of the overall impact on society. The COVID-19 vaccines, touted as highly effective (they are) and free (they are decidedly not), provide a striking example. Historically, vaccine development in the face of acute viral threats — like Zika and SARS — has not panned out financially for manufacturers, so, early in the coronavirus pandemic, they were not enthusiastic about developing and testing vaccines. Three months into the pandemic, as the gravity of the situation became impossible to ignore, the U.S. government launched Operation Warp Speed, an administrative mechanism to create financial incentives rich enough to motivate potential vaccine makers to develop, test, and manufacture vaccines quickly and in large quantity. Thankfully, this strategy helped produce highly effective vaccines in remarkably short order.

Early on, Operation Warp Speed spent five times more per person to procure vaccines for Americans than was spent by the European Union ($36 versus $7.25). This strategy was so successful that by mid-May 2021 the vaccination rate in the United States was far ahead of schedule and double that of the European Union. (By the end of July 2021, the EU had caught up to the United States, with about 70 percent of adults having received at least one dose of vaccine.) And given the oncoming toll in illness and death, in disruption of normal activities, and in economic loss, few would argue that we should have been more penny-wise with vaccine development and manufacture.

Not surprisingly, as the vaccines rolled out, the reputations of both Pfizer and Moderna—the first two manufacturers to be granted Emergency Use Authorization by the FDA—skyrocketed into the top ten among U.S. companies. Americans were grateful to them for working so quickly to integrate the best of medical science into effective vaccines, which allowed us to take the first step toward putting the pandemic behind us.

But a look behind the curtain reveals how the manufacturers exploited the public's desperation and hope, and how they adapted their tried-and-true profit-maximizing tactics to the entirely new opportunity presented by the pandemic.

First, the vaccine manufacturers' claim that the innovative drive of private enterprise was solely, or even primarily, responsible for the rapid development of COVID-19 vaccines was self-serving fiction. The foundational research that made the rapid development of vaccines possible had been completed in 2016 by scientists at the National Institutes of Health (NIH), working with researchers at Dartmouth College and Scripps Research Institute. They developed the technology to genetically engineer the exact sequence of amino acid building blocks that comprise the antibody-inducing spike proteins surrounding any specific strain of coronavirus.

In January 2020, armed with this technology, NIH researchers needed only a few days to turn the genetic code for COVID-19, provided by Chinese scientists, into a genetic "blueprint" for the vaccine. From that point it took Moderna only one month to develop and produce enough vaccine to begin large clinical trials. Zain Rizvi, a researcher with Public Citizen, summed up the mRNA vaccine manufacturers' role in the development of the COVID-19 vaccine succinctly: "Big Pharma started on third base and thought it hit a triple."

He was critiquing Pharma's attempt, through a PR campaign with a cautionary message, to leverage the reputational boost it received from producing successful vaccines. The campaign's headline: "An American Success Story We Should Never Take for Granted." Sponsored by the Pharmaceutical Research and Manufacturers of America, or PhRMA,* the message warned, "Unfortunately some in Congress want to enact partisan changes that could threaten access to medicines today and new treatments and cures in the future." In other words, if you dare cut the prices we are allowed to charge Medicare for our highest-revenue-generating drugs, we won't be able to continue to innovate. But this premise was in large part misleading. If Pharma really wanted to protect innovation, its warning to Americans should have been "Don't take *NIH research* for granted."

Second, despite paying top dollar for the initial round of vaccinations for Americans, Operation Warp Speed failed to leverage the federal government's generosity and purchasing power to ensure global vaccine equity. Although these agreements remain largely secret, the *New York Times* wrote, the U.S. government "used unusual contracts that omitted its right to take over intellectual property [which would have allowed it to provide vaccine for low- and middle-income countries] or influence the price and availability of vaccines."

At the $15- to $20-per-dose purchase price negotiated with the United States and other wealthy nations, Moderna projected more than $18 billion in sales in 2021 alone. With virtually the entire $1 billion cost of the research and development for its vaccine having been paid by the U.S. government (except for a generous $1 million donated by Dolly Parton) and the actual cost of production estimated to be as low as $3 per shot, it's not surprising that the price of Moderna's stock shot up elevenfold from January 2020 through May 2021.

Global sales of the Pfizer/BioNTech vaccine are expected to be even higher, $33.5 billion in 2021, making it by far the best-selling

---

* Three labels are used throughout this book to refer to the major players in the pharmaceutical industry: "Big Pharma" refers to the large pharmaceutical companies, "Pharma" refers to *all* pharmaceutical companies, and "PhRMA" refers to the specific trade organization mentioned here, the Pharmaceutical Research and Manufacturers of America, which is made up of the country's leading biopharmaceutical researchers and biomedical companies.

drug in the world, with an estimated profit margin "in the high 20 percent range." Pfizer CEO Albert Bourla broadcast his intention for post-pandemic pricing of the Pfizer vaccine in an interview with *Time*, referring to the *initial* vaccine sales to Operation Warp Speed and other wealthy nations as generating "a very, very marginal profit at this stage." Most people would consider a profit margin of almost 30 percent on billions of dollars of guaranteed sales to be considerably above the "very, very marginal" level.

Pfizer CFO Frank D'Amelio confirmed these expectations, explaining that the transition from acute COVID-19 pandemic to ongoing endemic would provide Pfizer with "a significant opportunity . . . from a pricing perspective," especially with the likely need for booster shots. Sales of Pfizer's COVID-19 vaccine are projected to reach almost $100 billion over the next five years. An analyst with the investment bank SVB Leerink estimated that Pfizer's profit margin on these sales will be a whopping 60 to 80 percent.

So, COVID-19 vaccines have not been offered free of cost. The vaccine manufacturers have been able to capitalize on the worst public health crisis in over a century by not charging consumers directly for the vaccines but rather extracting outsized profits from the taxpayers of wealthy nations. During the first fifteen months of the pandemic, nine new vaccine billionaires — including the CEOs of Moderna and of Pfizer's partner, BioNTech — gained $19.3 billion in personal wealth. In addition, eight already established billionaires with large investments in vaccines gained $32.2 billion. The total wealth of vaccine billionaires had increased by over $50 billion.

Third and lastly, the mRNA vaccine makers made sure they would squeeze out the most profit possible from the revolutionary new technology. Indeed, the best way for the manufacturers to maximize their financial gains was to ignore global needs and sell the largest possible share of their available vaccines for top dollar to wealthy nations while maintaining tight control of their patents and other intellectual property. Predictably, fifteen months into the pandemic 85 percent of the approximately two billion doses of COVID-19 vaccine administered worldwide had been given to residents of wealthy nations. At the other end of the spectrum, a mere 0.3 percent had been administered to people living in low-income countries. Dr. Tedros Ghebreyesus, the director general of the World Health Organization (WHO), called this "a scandalous inequity that is perpetuating the pandemic."

The WHO had tried to preempt this global inequity by launching the COVID-19 Technology Access Pool (C-TAP) in May 2020. The goal of C-TAP was to equip and train independent vaccine manufacturers in Latin America, Asia, and Africa by providing access to the patents *and* the technical know-how required to produce their own vaccines. Such a program, if it had been successful, would have at least partially freed less wealthy countries from being dependent on the voluntary largesse of wealthy countries and vaccine makers. When President-Elect Joe Biden's designated chief medical adviser, Dr. Anthony Fauci, was asked in January 2021 if he would encourage the United States to participate in C-TAP's sharing of technology, his response was emphatic: "That's an easy answer: yes, yes, yes." Pfizer CEO Bourla was equally emphatic, but in the opposite direction: "At this point in time, I think it's nonsense, and . . . it's also dangerous."

Moderna also rejected C-TAP, but much less forthrightly. In October 2020 the company issued a statement promising to act as responsible global citizens: "We feel a special obligation under the current circumstances to use our resources to bring this pandemic to an end as quickly as possible. Accordingly, while the pandemic continues, Moderna will not enforce our COVID-19 related patents against those making vaccines intended to combat the pandemic." Great talk, but the *Washington Post* reported that as of March 2021, Moderna had "taken no steps to share information about the vaccine's design or manufacture, citing commercial interests in the underlying technology." Without Moderna providing the requisite know-how, the company's offering of access to its patents did nothing to get vaccines to those living in low-income countries.

In May 2021, leaders from the International Monetary Fund, the

World Health Organization, the World Bank, and the World Trade Organization sounded a global all-hands-on-deck alarm, calling attention to the enormous impending negative consequences of not immediately increasing vaccination rates in low- and middle-income countries. They explained that on top of the imminent health risk to poor countries from high infection rates — by June 2021, 85 percent of the world's COVID-19 related deaths were occurring in developing countries — vaccine inequality would allow "deadly variants to emerge and ricochet back across the world." In other words, no country would be safe from COVID-19 until all countries were safe.

The leaders from the four organizations also explained that the global economy would sacrifice $9 *trillion* in lost productivity unless a 40 percent vaccination rate in all countries was achieved by the end of 2021 and "at least 60 percent by the first half of 2022." Most important, they concluded these goals could be achieved and economic disaster averted with an urgent infusion of about $50 *billion* to purchase vaccines for developing countries. (Coincidentally, this is exactly the amount of wealth gained by the seventeen vaccine billionaires during the first fifteen months of the pandemic.) The managing director of the IMF, Kristalina Georgieva, commented that this investment in global health would probably provide "the highest return on public investment in modern history."

Even in the most grotesquely selfish terms, an investment by the wealthy nations of $50 billion now — meaning in the early summer of 2021 — would return to them an estimated $1 trillion in increased tax revenues. Yet the slow walk by wealthy nations and vaccine manufacturers toward adequate rates of vaccination in nonwealthy countries — whether done on purpose or not — will pretty much ensure that the pandemic will continue and that booster shots will be necessary in wealthy nations in order to protect against the viral mutations breeding in under-vaccinated countries.

Moderna had publicly promised not to enforce its patents but failed to provide the additional technical know-how required to turn that promise into vaccine-making capacity. Pfizer CEO Bourla had promised that nonwealthy countries would "have the same access as the rest of the world" to its vaccine, but according to the World Health Organization, Pfizer also failed to follow through. As the vaccine maker was basking in the glow of its vaunted but false altruism, Richard Kozul-Wright of the United Nations Conference on Trade and Development commented that despite feigning commitment to low-income countries, Pfizer had prioritized sales to wealthy ones, making Pfizer's dramatic reputational turn-around "one of the great public relations triumphs of recent corporate history." But Big Pharma's triumph, like so many of its other triumphs, came at the public's expense.

The global public-health crisis caused by COVID-19 has provided a rare opportunity to observe the true extent of the vaccine makers' commitment to public responsibility and to private profit-seeking. Their willingness to ignore the needs of low- and middle-income countries in the interest of maximizing their own bottom line — despite the obvious risk to the health and economic well-being of wealthy nations — will make it easier to comprehend the "business as usual" situations and profit-making strategies described in this book. They are often quite jarring. And the example of the pandemic also helps show why, when the drug companies' profiteering techniques are focused on Americans, our health declines, and we are unable to provide all Americans with adequate health-care coverage, despite our off-the-charts spending.

I did not set out in my career to become a critic of the drug companies. After completing a family medicine internship, I served two years in the National Health Service Corps as a primary care physician in rural West Virginia. I then went back into training to complete two years of a family medicine residency and a two-year Robert Wood Johnson Fellowship, studying statistics, research design, and epidemiology. Although I had trained for a career of teaching, research, and practice in an academic medical center, I decided that my true calling was serving as a community-based family doctor.

When I entered private practice in 1982, I was confident that the care provided by American doctors (including me) was as good as the care available anywhere in the world. At that time, U.S. health-care costs were only slightly higher than those of other wealthy countries, and the U.S. death rate was well below the average in those same countries. I practiced family medicine for twenty years in a small town an hour north of Boston (and yes, I made house calls). I became the chair of family medicine at Lahey Clinic in Burlington, Massachusetts, and I have served on the faculty of Harvard Medical School since 1997, first as an instructor teaching primary care and, since 2010, as a lecturer in the Department of Health Care Policy.

During my years as a family doctor, I saw major changes in the practice of medicine. Beginning in the early 1990s, I became increasingly aware of the commercialism that was creeping into the sources of information I had been taught to trust — respected medical journals, educational lectures, and conferences. In 2001, drawing on the skills I had acquired during my fellowship twenty years earlier, I found serious and life-threatening discrepancies between the supposedly trustworthy scientific evidence published in the world's most respected medical journals and the actual data that drugmakers submitted to the FDA as summarized in FDA officers' reports, which had just started to be posted on the internet.

My curiosity was piqued by evidence that suggested an undisclosed increased risk of heart attack and stroke associated with the most heavily advertised drug at the time, the pain reliever Vioxx. By following a footnote in a medical journal article to its source, I found my way to a trove of data on the FDA's website providing undeniable evidence that cardiovascular risks were significantly higher with Vioxx than with an equally effective over-the-counter pain reliever, naproxen (brand name Aleve). But these risks had not been reported in the *New England Journal of Medicine* article that claimed Vioxx provided a safety *advantage* over the older anti-inflammatory drug, and the misleading advertisements led my own patients to request — and even demand — that I prescribe Vioxx for them. At that point I felt the need to understand how the integrity of the medical information delivered to doctors and patients was being distorted by the drug companies.

So in 2002 I left practice to write *Overdo$ed America: The Broken Promise of American Medicine,* to explain from a family doctor's perspective the extent to which this growing commercial intrusion was undermining medical care. Three days before the book was published, the *New York Times* ran an op-ed I wrote, titled "Information Is the Best Medicine," in which I argued that doctors needed better access to clinical trial data (I specifically cited data showing the significantly increased cardiovascular risk of Vioxx) to provide their patients with safe and effective care. *Overdo$ed America* was published on September 21, 2004. Nine days later Merck abruptly pulled Vioxx off the market, not because of my book but because the results of yet another study had documented the increased risk of heart attack and stroke associated with the drug. It was the biggest drug recall in U.S. history.

I immediately sat down at my computer to write another op-ed, explaining why Merck's sudden withdrawal of Vioxx ought not to have come as a surprise. About half an hour later, my publicist called to tell me that a limousine would be picking me up in fifty minutes; I was to do three live national television interviews by satellite from Boston and then fly to New York City, where Katie Couric would interview me on *Today* the following morning. At this point I had done a total of one television interview on a local station, so the agenda for the next twenty-four hours was, to say the least, intimidating.

When I arrived in the greenroom of the *Today* show, I saw pictures of the hosts lined up on one wall. I had to ask my publicist which one was Katie. I was taken to the set and seated on a couch, where I waited nervously. Katie finished up another segment and came over and joined me, and we were quickly (and comfortably) in the middle of a conversation, with millions of Americans watching on live TV. I was invited back six weeks later, when another study showed that Celebrex too might increase the risk of cardiovascular problems (although this turned out not to be a major issue, and the drug was not withdrawn from the market).

Soon lawyers coordinating national litigation involving thousands, and sometimes tens of thousands, of plaintiffs allegedly injured by one or another prescription drug began to ask me to serve as an expert witness. This work turned out to be the most challenging and consequential I had ever undertaken (besides caring for individual patients at critical moments). After signing a confidentiality agreement (without which, access to the confidential corporate documents produced in discovery was not allowed), I would begin my investigation. Typically, the first step was reviewing key documents, culled by plaintiffs' lawyers, which were deemed to be evidence that the manufacturer had misled doctors, the public, and insurers by exaggerating the benefit of a drug or minimizing the harm it might cause. But that was just the starting point.

I was granted access to the computer files of the relevant drug-company executives and scientists. No longer on the outside, wondering whether the company-sponsored science really supported the claims being made about a drug, I had access to millions of confidential documents — scientific data, e-mails, business and marketing plans, internal slide presentations about science and marketing, and so on. With this access I could piece together what the science really showed and evaluate:

- whether clinical trials had been designed to produce results that were misleadingly advantageous to drug sales;
- whether corporate marketing and business plans designed to capitalize on marketing research called for misrepresentation of scientific evidence;
- whether actual scientific data from clinical trials, when analyzed according to the rules the manufacturer had established before the study was started, supported the results presented in medical journals and marketing materials;* and
- whether marketing plans and slides developed by the manu-

---

* I often worked with a statistician who was also serving as a plaintiffs' expert.

facturer to "educate" doctors about its drug accurately presented the scientific findings.

Thus I could compare, on the one hand, what doctors would reasonably conclude about the safety and efficacy of the drug based on the information provided by the manufacturer with, on the other hand, what doctors would likely conclude if they had accurate and unbiased summaries of *all the information*. Or, stated more succinctly, I could determine if the reported benefits and risks of a given medication represented the best available scientific evidence. Often they did not.

Vioxx was the first litigation I worked on, but over the next ten years I served as an expert in about fifteen other cases and was deposed by many lawyers hired to defend many different drugmakers and one medical-device maker. These cases were all civil litigation: Plaintiffs' attorneys sought to win compensation for individuals who had allegedly suffered personal injury or for institutions — such as union health plans, insurers, or governmental bodies — that had allegedly suffered economic injury arising from fraudulent claims of a drug's efficacy, safety, or value.

After I had thoroughly reviewed the information available in the company's files and had written an expert report, I would be deposed by the lawyers hired to defend the drugmakers. These depositions often felt like heavyweight boxing matches, as accomplished and tough corporate attorneys, informed by their own scientific experts and backed by teams of lawyers and support staff, challenged my analyses. My reports and opinions had to be rock solid; any weakness would be quickly and embarrassingly exposed.

In addition to serving as an expert in civil litigation, I was able to bring some of what I learned to the FBI and the U.S. Department of Justice as an unpaid consultant. In one of these cases, Pfizer pleaded guilty to a felony for marketing its arthritis drug Bextra (a cousin of Vioxx) "with the intent to defraud or mislead" and, according to a DOJ press release, agreed to pay "a criminal fine of $1.195 billion, the largest criminal fine ever imposed in the United States for any matter." Even so, nobody went to jail for committing this felony, and the documents, including those I brought to the DOJ and my analyses showing how Pfizer had allegedly misled doctors, remain sealed under the terms of the settlement. I am still not at liberty to explain how Pfizer hoodwinked (or, more accurately, feloniously misled) American doctors into prescribing enough Bextra to make the company a handsome profit.

The relationship between the science and the marketing of almost all the drugs I investigated turned out to be a variation on this theme. I saw that doctors were being misled by the drug companies —misled about the results of clinical trials and misled about how the claimed findings should be integrated into optimal care. Where the terms of the settlement allow, I have included in this book the findings of some of the cases I worked on. And after seeing this pattern repeated again and again, I realized how much important information was being concealed and saw how the drug companies manipulated the scientific evidence that health-care professionals rely on.

This experience gave me a unique window into the dysfunction of American health care. Neither the public nor the nation's physicians are aware of the extent to which drug companies and other commercial interests produce and control the information that guides medical decisions. This is the key to understanding how America can be spending so much more money on health care than the other wealthy nations do while the health of its citizens continues to fall farther and farther behind.

Each of the chapters in part I discusses a drug or class of drugs that became far more widely used than the science justified or good care warranted. These examples reveal the range of tactics drug companies use to oversell their drugs without regard for the consequences. They range from manipulating medical journals, including some of the most influential in the world, to illegally marketing drugs (I testified in a trial that found a global pharmaceutical company had committed fraud and engaged in a racketeering conspiracy), to recommending, in treatment guidelines, that more than half of U.S. adults between the ages of forty and seventy-five take statins to lower cholesterol — though 60 percent have no history of heart attack, stroke, or diabetes, to employing deception to promote the use of unconscionably expensive insulin for people with type 2 diabetes when far less expensive insulin would have been at least as safe and effective.

Part II shifts from these specific examples of overprescribed drugs to a broader discussion of how the changes in American society over the past four decades have allowed commercial interests to control much of the medical knowledge that now guides our health care. Beginning around 1980, the primary mission of for-profit, publicly held corporations in the United States began a fundamental transition: Companies moved from addressing the needs of a broad array of stakeholders — including customers, employees, communities, and shareholders — to focusing narrowly on maximizing financial return to shareholders. This transition, along with opening the money spigots from corporations and superwealthy individuals, created the political and economic context that led to unprecedented inequality of wealth and the declining health of Americans relative to the citizens of other wealthy nations.

In this corporate-profit-maximizing context, clinical research became ever more privatized, focusing on the creation of wealth rather than health. Universities and academic medical centers needed to seek funding from drug companies to continue their research programs, and financial ties with those same drug companies — something that was once beneath the dignity of academic physicians — became a sign of prestige. Drug companies gained control of doctors' long-trusted sources of medical knowledge and then skillfully exploited that trust.

But perhaps most important, the social contract ensuring that for-profit corporations served their customers, employees, *and* communities was abandoned. The revised financial incentives evolved into an interconnected system that now produces and disseminates so-called scientific evidence that, first and foremost, serves the financial interests of the health-care industry and its investors. Thus, the dysfunction in American health care is in large part the result of rational—but antisocial—behavior by executives of drug and device companies, which then "trickles down" to academic medical centers and researchers, medical journals, medical societies, and respected nonprofit organizations. In other words, the way we now produce the knowledge that directs our health care is in a state of market failure. The primary actors are thriving, but society as a whole is being harmed.

The final part of this book begins with a summary of the achievements and limitations of recent reform, particularly Obamacare. Going forward, meaningful reform — expanding health-care access to all Americans, improving Americans' health, and containing our medical spending—will require standing up to the hyper-profitability of the vested interests (Big Pharma, insurance companies, and hospitals). This will be a very steep hill to climb, one that will call for a new constituency for health-care reform led by a coalition of health-care professionals, purchasers (including non-health-care-related businesses, unions, and government), and consumers. But change will not happen until the constituency for reform becomes a more powerful force than the corporate money and lobbyists that currently hold sway over our politicians. This is the only way to bring Americans' health *up* to the level enjoyed in other wealthy nations and the cost of that health care *down* to the level in those same nations.

The illusion of American medical exceptionalism must yield to the basic principles of good science. And we must dispel the misconception that the least regulated market is the most efficient, which means allowing government intervention, when needed, to make sure that the health-care market effectively serves the interests of society. Our goal must be to find a better balance between commercial profit and public benefit.

And, daunting though it may be, I believe there is a path forward.

PART I

Health Care American-Style

VIOXX: AN AMERICAN TRAGEDY

Don't believe anything, not one thing, put out by a pharmaceutical company. Just don't believe it. You start from there.

— CATHERINE DEANGELIS, editor in chief,
Journal of the American Medical Association

L ate in 2004, I received a heartbreaking note from a woman
who had seen me on television discussing safety issues associated with Vioxx and Celebrex, two widely used pain reliev-
ers. Mrs. Susan Palmer wrote to ask if I would be willing to review
her daughter's medical records: "It would be a blessing to us. Please
help us. My daughter, Stacey, was 17 years old."*

Several months earlier, Stacey, a healthy honors student and
budding gymnastics star, fell and hit her head during practice.
She blacked out briefly, and when she came to, she had pain in her
right wrist. Stacey's mother picked her up at school and brought
her to the emergency room. There was no evidence of serious head
trauma, but wrist X-rays revealed a fracture. A cast was put on, and
she was instructed to take ibuprofen (Motrin or Advil) as needed
for pain and to follow up with a pediatric orthopedist.

---

* Details have been changed to protect anonymity.

About a week later Stacey began having headaches. Ibuprofen provided little relief, so Stacey's mother brought her to their family doctor. The doctor found she had some tenderness in the back of her neck and the right side of her head, but Stacey's examination was otherwise normal. A CT scan of her head was also normal. Because ibuprofen was not working, the doctor gave Stacey eight Vioxx 25 mg tablets, samples provided by the manufacturer, to be taken once daily, and made a follow-up visit for one week later. Vioxx, a pain reliever marketed as a safer version of ibuprofen and similar drugs, had recently been approved to treat migraine headaches.

That Stacey's family doctor had samples of Vioxx meant a Merck drug rep had "detailed" him (that is, given him information about the company's new drug, trying to persuade him to prescribe it). The rep probably had visited the office, brought food and small gifts (like pens and notepads emblazoned with vioxx) for the doctor and his staff, and "educated" them about the benefits of Vioxx. And the rep probably left marketing materials, possibly including a re-print of the VIGOR (short for Vioxx Gastrointestinal Outcomes Research) article that had been published in one of the most influential medical journals, the *New England Journal of Medicine* (*NEJM*), on November 23, 2000. The article concluded that Vioxx had a safety advantage over an older anti-inflammatory drug, naproxen (Aleve), because it caused fewer serious stomach complications.

But when Stacey returned for a recheck, she reported that her headaches were occurring daily and had become so severe that she was going to the school nurse's office two or three times a week to lie down. She also reported having blacked out while walking to school. Again, Stacey's exam was normal. The doctor suspected post-concussion syndrome.* Because the Vioxx had not helped, he advised Stacey to go back on ibuprofen at a higher dose and ar-ranged for her to see a neurologist. Stacey never got to that appoint-

---

* Typically headache and dizziness that begin within a week or so after a head trauma and resolve within several months; the symptoms are easily mistaken for tension headaches.

ment. Four days later, at 4:30 a.m., Mrs. Palmer found her daughter dead in her bed, the victim of a massive stroke.

This needless tragedy almost certainly would not have happened if it weren't for the failure of three guardians of public safety: the manufacturer, which concealed the risk of Vioxx; the *New England Journal of Medicine*, which published the study claiming the drug was safe and failed to correct the article after its editors became aware of the undisclosed risk; and the FDA, which failed to act effectively as soon as it became aware of that risk.

DOCTORS WERE FALSELY REASSURED

As the Palmer family struggled with their loss, one of the world's largest drug companies was dealing with a crisis of its own. On September 30, 2004, less than three weeks after Stacey died, Merck pulled its blockbuster Vioxx off the market. A just-completed Merck-sponsored study had shown 25 milligrams of Vioxx per day — the dose Stacey had taken — doubled the risk of heart attack and stroke. According to Dr. Peter Kim, the president of Merck Research Labs, the study showed an "unexpected increase in cardiovascular disease rates." He told the *New York Times,* "What we saw was stunning."

As a Merck executive, Dr. Kim had good reason to be unhappy with the latest study results — but he should not have been surprised. More than four years earlier, as shown by internal documents, Merck executives lamented similar results from the VIGOR study. On March 9, 2000, when the VIGOR data first became available to Merck scientists, Merck's then head of research, Dr. Edward Scolnick, wrote in an internal e-mail that the cardiovascular dangers of Vioxx (meaning increased risk of heart attack, stroke, and blood clots) were "clearly there" and "a shame." Nonetheless, his e-mail concluded reassuringly: the class of drugs of which Vioxx was a member (the COX-2 inhibitors) "will do well *and so will we*" (italics mine).

The VIGOR data Merck sent to the FDA in August 2000 — by law, this had to be a full accounting of the trial's results — stand in stark contrast to the reassurance about the safety of Vioxx claimed in the *NEJM* article published on November 23, 2000. The article did report that people who were treated with Vioxx had four times as many heart attacks as those treated with another anti-inflammatory drug (naproxen) but then stated that the increase occurred in people who should have been taking aspirin to prevent heart attack and stroke but were not allowed to by the study design. Importantly, of the thirteen authors, eleven were researchers with financial ties to Merck, and two were Merck employees, both of whom had received Dr. Scolnick's e-mail of March 9, 2000. The Merck-employed authors knew, or should have known, that the *NEJM* article they coauthored had underreported the increased risk of heart attack associated with Vioxx.

When the 2000 article presenting the results of the VIGOR trial was first published, I read it with interest. However, it was a second *NEJM* article, published the second week of August 2001, that created a watershed moment in my career. I was eating lunch and reading medical journals in my office when I came upon an *NEJM* overview of the COX-2 inhibitors Vioxx and Celebrex. The article claimed (incorrectly it turned out) that these drugs caused "significantly fewer serious gastrointestinal adverse events" than older anti-inflammatory drugs. The review did, however, include more detail about the cardiovascular events in the VIGOR trial than the initial *NEJM* article had, reporting that people treated with Vioxx experienced significantly more heart attacks and strokes than those treated with naproxen.* But the article then dismissed this increased danger with a peculiar argument: "The difference in major cardiovascular events in the VIGOR trial *may reflect the play of chance*. The end point was prespecified, and the difference in the fre-

---

* From the article: "The rates of nonfatal myocardial infarction, nonfatal stroke, and death from any vascular event were higher in the rofecoxib [Vioxx] group than in the naproxen group (0.8 percent vs. 0.4 percent, P<0.05)."

quency of events was statistically significant, but the absolute number of cardiovascular events was small (*less than 70*)" (italics mine).

The moment I read those two sentences, I knew something was dreadfully wrong with the information presented about Vioxx in the world's most influential medical journal. First, even a beginner knows statistical significance is based on the number of events that occurred, so dismissal of statistical significance because of a small number of events made no sense and reeked of bias. Second, what a monumental double standard — dismissing the cardiovascular harm because "less than 70" events had occurred but at the same time touting the reduction in serious gastrointestinal events associated with Vioxx, though there were even fewer of those. Nobody on the Merck side of this issue was blowing off the gastrointestinal (GI) benefit of Vioxx because only 53 serious GI events had occurred in the VIGOR trial.

The inexplicably favorable treatment of Vioxx in this second article made me wonder if either of its two authors had financial ties to Merck. I wasn't surprised when I read the fine print: Both did. But I was puzzled because at the time, the *NEJM* had a clear policy that authors of review articles and editorials were expected not to "have any financial interest" in the companies that made the products discussed. That the *NEJM* editors changed this policy in June 2002 to allow such conflicts of interest does not minimize their wanton disregard of their own rules in 2001.

The data presented in the review article showed that the authors had access to more VIGOR data than had been published in the 2000 *NEJM* article. I tried and failed to find the source of their information. But just two weeks later, an article in the *Journal of the American Medical Association (JAMA)* reported even more of the VIGOR data; it showed that the risk of suffering a cardiovascular event was 2.38 times greater with Vioxx than with naproxen. I pasted the web address from footnote 20 of the *JAMA* article into a web browser and was suddenly looking at the FDA's assessment of the data Merck had submitted from the VIGOR study.

As I read through the FDA reports dated February 2001, it became patently clear that the *NEJM* articles' assurances about the safety of Vioxx were based on three brazenly misleading claims. First, the authors simply omitted three heart attacks that occurred in people treated with Vioxx. They cooked the books, lowering the number just enough to avoid a statistical threshold that would have shown that, aspirin or no aspirin, Vioxx was triggering heart attacks.* Second, the original *NEJM* article reported heart attacks alone, rather than all cardiovascular events (heart attacks, strokes, and blood clots); if it had included *all* cardiovascular events, as called for in the study protocol, it would have shown a more than double (and statistically significant) increase in cardiovascular risk associated with Vioxx. And in a third, equally audacious breach of scientific integrity, both *NEJM* articles simply failed to report that people treated with Vioxx developed 21 percent more serious illnesses than those treated with naproxen. In other words, the overarching finding from the VIGOR study was that Vioxx was significantly *more dangerous* than naproxen and provided no better pain relief.

A fundamental principle drummed into me during my medical training was that a doctor's practice should be guided by the findings of high-quality studies published in peer-reviewed medical journals, but I suddenly understood that at least some information in the *NEJM* could not be trusted. Now I wondered if this was also true for other journals. I returned to the FDA website, searching for the data behind an article I'd read in the *Washington Post*. The newspaper had reported that a *JAMA* article claiming to show that Celebrex caused fewer serious GI complications than older over-the-counter anti-inflammatory drugs was also based on

---

* When those three omitted heart attacks were added back in, the results showed a significantly increased risk of myocardial infarction for all patients in the VIGOR study. The *NEJM* Expression of Concern, published in December 2005, stated: "Lack of inclusion of the three events resulted in . . . the misleading conclusion that there was a difference in the risk of myocardial infarction between the aspirin-indicated and aspirin-not-indicated groups." But this was far too little, far too late.

a misrepresentation. It turned out the *JAMA* article had relied on only six months of data from the large Pfizer-sponsored Celecoxib Long-Term Arthritis Safety Study (CLASS). However, CLASS was a *twelve*-month study. At the end of twelve months, patients who took Celebrex did *not* have significantly fewer serious GI complications than those who took older anti-inflammatory drugs.

The two most trusted medical journals in the United States had published incorrect, manufacturer-biased reports about major drugs. The FDA knew that both articles had misrepresented the data but did not correct the misleading information.

I was stunned.

THE *NEJM* WAS PART OF THE PROBLEM

A Seattle pharmacist, Jennifer Hrachovec, PharmD, had found the FDA data about Vioxx and the VIGOR trial at least two months before I did. She and a physician colleague submitted a letter to the *NEJM* in June 2001 pointing out that the journal's report had not included all the heart attacks that occurred among people treated with Vioxx, and thus the published results made Vioxx look less dangerous than it really was. The *NEJM* responded that space was limited and did not publish the letter.

Two months later, in August 2001, Dr. Hrachovec called in to a radio show on which Dr. Jeffrey M. Drazen, the editor in chief of *NEJM*, was a guest. According to the *Wall Street Journal*, Dr. Hrachovec "begged" Dr. Drazen to update the article: "My concern is that doctors are still using [Vioxx] and exposing their patients to higher risks of heart problems and they just don't even know that that's the case." (Remember, this was three years *before* Stacey Palmer died.) Dr. Drazen responded dismissively, "We can't be in the business of policing every bit of data we put out." Indeed, as we will see, the *NEJM* was *not* in that business.

On September 17, 2001, one month after Dr. Hrachovec's call to the radio show, the FDA sent an official warning letter to Merck, stating the company's marketing of Vioxx minimized "the potentially serious cardiovascular findings" observed in the VIGOR trial and was "false, lacking in fair balance, or otherwise misleading." In uncharacteristically forthright language, the FDA letter called a Merck press release that claimed Vioxx had a favorable cardiovascular safety profile "simply incomprehensible given the rate of [heart attacks] and serious cardiovascular events compared to naproxen." Finally, based on "the fact that your violative promotion of Vioxx has continued despite our prior written notification," the FDA demanded Merck "immediately [cease] all violative promotional activities and dissemination of violative promotional materials for Vioxx." Strong words from the FDA. The letter received some coverage in the press and was available on the FDA website (if you knew where to look). But still, this critically important information did not come from a source that doctors considered more reliable than their medical journals, and an FDA official later testified to the Senate that the FDA did not do enough to make sure doctors understood the cardiovascular risk of Vioxx.

Even though the FDA officials grasped the seriousness of this problem, they left a loophole big enough to drive tens of thousands of funeral processions through: They did not stop Merck from continuing to purchase hundreds of thousands of reprints of the incomplete and falsely reassuring *NEJM* article and hand them out to doctors. These reprints repeated the very same marketing claims that the FDA warning letter had expressly forbidden Merck to make, but the company let the *NEJM* get paid to do the talking. And the *NEJM* was apparently happy to do so, selling more than 929,000 reprints of its misleading article (mostly to Merck) — far more than enough to give one to every prescription-writing doctor in the United States — and bringing between $697,000 and $836,000 into the journal's coffers.

When Vioxx was pulled off the market in 2004, it was the biggest drug recall in history. Twenty-five million Americans had taken the drug during the previous five years. Patients were finally told that the drug that had been relieving their aches and pains (although, according to Merck's own clinical trials, no more effectively than older, far less expensive anti-inflammatory drugs) was more than doubling their risk of heart attack and stroke. And just eighteen days earlier, Stacey Palmer had died.

HARVARD MEDICAL STUDENTS WANTED THE TRUTH; IT WAS NOT AVAILABLE FROM THE *NEJM*

Soon after Merck rocked the medical world with its abrupt withdrawal of Vioxx, a group of Harvard Medical School students invited me to give a lunchtime lecture explaining what had happened. By the time Professor Byron Good, chair of the Department of Social Medicine, called the room to order, it was packed. (Free lunch may have helped.) People were sitting on the floor and standing against the wall at the back of the room.

I began by reviewing the elegant science behind the COX-2 hypothesis. Older anti-inflammatory drugs such as ibuprofen and naproxen work by suppressing what was thought to be a single enzyme, cyclooxygenase (COX), which catalyzes a cascade of events leading to inflammation. These drugs are quite effective at reducing pain and inflammation but have the unfortunate tendency to cause stomach symptoms and, occasionally, serious GI complications like bleeding or perforated ulcers.

In the early 1990s, scientists discovered there were two different forms of cyclooxygenase. COX-1 protects the lining of the stomach; COX-2 leads to the inflammatory response. Ibuprofen and naproxen are called "nonselective" because they block both forms of cyclooxygenase. Vioxx and Celebrex selectively block primarily the COX-2 enzyme, and the hope was that they could decrease inflammation and pain without causing as many serious stomach problems.

However, human biology is rarely that simple. It turns out that COX-1 also makes platelets stickier, which increases the risk of developing blood clots. COX-2 exerts the opposite effect—it opens blood vessels and makes platelets less prone to clump together, thus decreasing the likelihood of developing blood clots. So, while selective COX-2 inhibitors were theoretically gentler on the stomach, they reduced the clot-inhibiting activity of COX-2 without reducing the clot-promoting activity of COX-1, thereby *increasing* the risk of heart attack, stroke, and peripheral blood clots.

I then showed the students excerpts from a *Wall Street Journal* article headlined "E-Mails Suggest Merck Knew Vioxx's Dangers at Early Stage." Internal Merck e-mails from as early as 1996 show Merck was aware that allowing people in a Vioxx clinical trial to take low-dose aspirin could mask the drug's hoped-for GI benefit, but not allowing aspirin could increase the likelihood of cardiovascular events. In 1997 Merck's vice president for clinical research (and an author of the VIGOR study manuscript) suggested that a way out of this bind was to exclude people from the VIGOR study if they had a history of cardiovascular disease and thereby eliminate people who were taking low-dose aspirin. And this is exactly how the VIGOR trial was designed.

I presented slides from the FDA review of the VIGOR study, showing Vioxx more than doubled the risk of heart attack, stroke, and blood clots and was significantly more dangerous overall than naproxen. Then I posed the key question: If Merck's data showed Vioxx was more dangerous and no better for pain than far less expensive over-the-counter drugs, why had so many doctors prescribed it? At this point I showed my audience the falsely reassuring findings of the VIGOR study and the review article on Vioxx and Celebrex published in the *NEJM*. Good doctors, after all, try to practice evidence-based medicine, and they expect what they read in peer-reviewed medical journals to provide exactly that.

I finished my talk by summarizing an article published in *The Lancet* the previous week by Dr. David Graham of the FDA and others. The authors estimated that Vioxx was responsible for between 88,000 and 140,000 heart attacks during its five years on the U.S. market. This meant 40,000 to 60,000 Americans died as a result of taking Vioxx, roughly the number of American soldiers who died in the Vietnam War.

AN ATTEMPT AT INTIMIDATION

A couple of weeks later, the Harvard medical students invited me back, this time to discuss the 2004 update to the guidelines for prescribing statins (to be discussed in chapter 3). We were meeting in that same horseshoe-shaped room; a few minutes before Professor Good introduced me, I noticed two older gentlemen sitting unobtrusively in the back left corner. I approached them, introduced myself, and asked what brought them there. They identified themselves as Gregory Curfman, the executive editor, and Stephen Morrisey, the managing editor, of the *New England Journal of Medicine*.

Under normal circumstances I would have been honored by the attendance of two *NEJM* editors. But following so closely upon my previous talk, their presence made me apprehensive. I knew that the journal was still not being forthright about the increased cardiovascular risk associated with Vioxx in the VIGOR trial. And the presence of the *NEJM* editors at my second lunchtime lecture was likely an indication of their concern about my willingness to openly discuss the extent to which they were not being forthright.

A few weeks after my second talk, my phone rang; the caller ID read *NEJM*. The caller identified herself as editor in chief Dr. Jeffrey Drazen's administrative assistant and informed me she was aware I would be completing a lecture to a Harvard Medical School–sponsored Continuing Medical Education program at 11 a.m. on Wednesday, March 23, 2005. She then announced I was to be at Dr. Drazen's office immediately following the lecture. Taken aback, I responded a bit tartly, saying the attendees usually wanted to continue the discussion, so the earliest I could be there was 11:30. That was acceptable, she said.

After hanging up the phone, I didn't know whether to feel complimented or intimidated — or both. Jeff Drazen was one of the most prestigious and powerful doctors in the world. I wondered if he was going to suggest that I write a piece for the journal describing the hazards of Vioxx that had not been fully or accurately reported in its November 2000 article; perhaps he might even offer me the opportunity to write critical pieces periodically.

On the appointed day, after delivering the lecture and chatting with attendees, I walked across the medical school campus to the *NEJM*'s offices on the top floor of the Countway Library. Upon arrival, I was quickly ushered into Dr. Drazen's large office, which was tastefully decorated with cultural artifacts presumably from his travels. We shook hands and sat down at a circular table. Dr. Drazen began by telling me that he was feeling jet-lagged, having just returned from a trip to China. We chatted briefly about the growing need for current medical information there. Then, practically in midsentence, Dr. Drazen cut to the chase: "How do you think I feel when I am making hospital rounds with medical students, and they tell me you are criticizing the *New England Journal* in your lectures?" Since I wasn't absolutely sure how he felt (though it didn't look good), I didn't answer.

I was, however, guilty as charged. I had thought, evidently naively, that the fundamental purpose of academic institutions was to seek the truth. So I was gobsmacked by what seemed to be Dr. Drazen's position: that I should not have explained to medical students how the *NEJM* articles differed from Merck's own VIGOR study data; that the *NEJM* did not have an obligation to its physician readers to correct any misinformation reported in its pages; and (by implication) that the *Veritas* on the Harvard University shield actually meant that truth was conditional and subject to commercial considerations.

I tried to engage Dr. Drazen in a discussion about the difference between the data Merck had submitted to the FDA and the two *NEJM* articles about Vioxx. He seemed more concerned about my public criticism of the articles than about what the data really showed. Dr. Drazen then asked me another question: "If you were so upset, why didn't you write a letter to the editor?"

"Funny you should mention that," I responded, "because I recently wrote a letter to the *NEJM* about an article that advocated more widespread use of the test for C-reactive protein, and you didn't publish it."

Dr. Drazen replied, "We don't have enough room to publish all of the letters we receive."

"I understand," I said, "but you didn't publish *any* letters about that article." In addition, I hadn't found the FDA data showing the *NEJM* article was incomplete and misleading until August 2001 (when the *JAMA* article pointed me to the FDA reports), about six months after the period for which letters regarding the VIGOR article would have been considered.

Our conversation continued but did not progress. He wouldn't budge from his defense of the authority and reputation of the *NEJM,* and I was adamant the journal had a responsibility to correct the errors about Vioxx as soon as it became aware of them. Although I am not a mind reader, it felt to me like Dr. Drazen was attempting to use his power and position to bully me out of publicly criticizing the *NEJM*'s handling of its misleading articles that minimized the cardiovascular and overall risk of Vioxx. I realized it was time to stop our useless back-and-forth. My parting words to Dr. Drazen were presumptuous: "History will not treat you well."

So far, I have been wrong.

A month after my meeting with Dr. Drazen, Professor Jerry Avorn and I shared the podium for a Cabot Lecture* at Harvard Medical School. The lecture had an unwieldy title: "Vioxx, Bextra, and the Erosion of Trust in Drug Regulation and Our Clinical Knowledge." The lecture was in the Cannon Room, an amphitheater built in 1906 whose interior surely hadn't been updated since.

---

* A monthly lecture for Harvard Medical School faculty and students sponsored by the Division of Primary Care.

Professor Avorn spoke first, addressing the science of the selective COX-2 inhibitor arthritis drugs and the process of new drug evaluation. While Jerry was speaking, I scanned the audience and noticed that the two *NEJM* editors, Drs. Curfman and Morrisey, were once again sitting high up in the last row on the extreme left. This time, I assumed their purpose was either to intimidate me into tempering my comments about the *NEJM*'s role in the Vioxx disaster or to act as a real-time "truth squad." As Jerry was finishing, I was unsure if or how I should acknowledge the editors' presence.

I made my way to the podium still undecided. When I opened my mouth to speak, instead of my intended remarks, out came "I am going to talk about the erosion of trust in our clinical knowledge and in particular the role the *NEJM* played in the Vioxx debacle." Then, looking up into that dark corner of the amphitheater, I continued, "And we are fortunate to have two editors from the *NEJM* here. So if I say anything that is not correct, they can set the record straight." The thinly veiled intensity of my comment created an awkward moment.

I presented a shortened version of the lunchtime lecture I had given two months earlier, making no changes to avoid potential repercussions from my two guests in the back row. As I spoke, I sneaked an occasional look at them. They maintained a stolid silence, offering neither comment nor correction. (Only later did I consider that they might have been there to better understand my critique of the *NEJM* articles.)

THE *NEJM* ISSUES AN "EXPRESSION OF CONCERN"

On December 8, 2005, seemingly out of the blue and more than five years after its publication of the VIGOR trial results, the *NEJM* issued an Expression of Concern (EOC) — one step short of a correction — regarding the article. Omission of three heart attacks in the original article, the EOC asserted, had allowed the authors to understate the increased risk of heart attack associated with Vioxx. Well and good, but why would the *NEJM* correct the record (or, more accurately, "express concern") when Vioxx had already been off the market for more than a year?

The answer surfaced six months later, reported in detail in the same *Wall Street Journal* article that discussed the conversation between the Seattle pharmacist and Dr. Drazen. As it turned out, Merck was being sued in Houston's federal district court by the family of fifty-three-year-old Richard "Dickey" Irvin, who died from a heart attack after taking Vioxx. The true purpose of the EOC (and the reason for its timing) was revealed in a series of internal *NEJM* e-mails obtained by the newspaper. The first e-mail, written by an *NEJM* editor to the staff on December 7, the night before the EOC was issued, stated unabashedly: "The reason [for the EOC] is that tomorrow's testimony in the [Irvin] Vioxx trial may involve part of a deposition that Greg gave." "Greg" was Dr. Gregory Curfman, one of the two *NEJM* editors who had attended the two lectures described above; he had been deposed in November 2005 for the Irvin trial. According to the *Wall Street Journal*, he had acknowledged in his videotaped deposition that "lax editing [on the part of the *NEJM*] might have helped the authors make misleading claims in the [VIGOR] article."

NEJM public relations consultant Edward Cafasso e-mailed the *NEJM* editors later that night: "I believe that given what a public punching bag Merck has become, there is more than enough information and more than enough context in the statement to drive the media away from *NEJM* and toward the authors, Merck and plaintiff attorneys." On December 9, the day after the *NEJM*'s Expression of Concern was issued — and covered by the press — Cafasso wrote again, "The story is playing out exceptionally well."

To ensure that the story would continue to play out well, a list of talking points was e-mailed to *NEJM* editors on December 12,

2005. Among those points was one advising them to deny the Expression of Concern had anything to do with the trial in Houston. This was patently false — of course the EOC was related to the trial — but it clarifies the journal's priorities.

Sadly, those priorities did not include the safety of Stacey Palmer or all the other tens of thousands of Americans who had suffered cardiovascular complications as a result of taking Vioxx. Many of those injuries and deaths could have been prevented if the *NEJM* had corrected its erroneous article as soon as it knew of the problem; instead, it left the article uncorrected, allowing Merck to profit from the drug's blockbuster status for another three and a half years and the journal itself to profit from selling reprints of the misleading article to Merck.

WHEN MARKETING TRUMPS SCIENCE

Stacey's autopsy showed she had died from a massive stroke. Given Stacey's youth, the absence of medical problems that would have predisposed her to such a disastrous stroke, and the known doubling of cardiovascular risk associated with Vioxx, it is probable that a Vioxx-induced blood clot led to her demise. This was the likely proximate cause of Stacey's death. But the real cause was the simultaneous failure of those three supposed guardians of public safety mentioned earlier, whose collective negligence allowed Stacey's family doctor to reasonably believe that Vioxx was the best choice to treat the girl's post-concussion headaches.

Merck failed the Palmer family by not being forthright about the risk of heart attack and stroke associated with its blockbuster drug. The design of the VIGOR study — excluding people with a cardiovascular history who were taking low-dose aspirin — minimized the chance that the potential cardiovascular risk of Vioxx would be revealed. When that risk nonetheless became clear, it was simply not reported in the published results. Even after Merck received a formal FDA warning letter and included the increased risk of serious cardiovascular events associated with Vioxx in the 2002 update of its product label, it continued to purchase reprints of the misleadingly incomplete *NEJM* VIGOR article and to urge doctors to prescribe and patients to request Vioxx — spending more money to market a drug than any company in history.

The *NEJM* failed the Palmer family by not setting the record straight about its article of November 23, 2000, as soon as its editors became aware that it understated the cardiovascular risk of Vioxx. Further, the *NEJM* continued to sell reprints of the misleading article to Merck, so it could use the journal's good name to persuade doctors to prescribe a dangerous drug. The *NEJM* remained silent for more than four years — until the drug had been off the market for fourteen months — before it partially set the record straight, and even then, editors acted only because the journal's reputation was at risk.

And the FDA failed the Palmer family by not intervening in the unfolding public health emergency after its officers' reports, based on Merck's data, had clearly identified the risks. Testifying before the Senate in 2005, FDA official Dr. Sandra Kweder explained that the agency did not have the authority to impose changes to a drug's label; the manufacturer had to do that. But Dr. Kweder also acknowledged that the FDA was wrong in failing to make sure the information about the increased cardiovascular risk of Vioxx was "in the forefront of the consciousness of the prescribing physician." Had Stacey's family physician been fully informed of the increased risk of heart attack and stroke associated with Vioxx instead of — most likely — having been actively misinformed about those grave risks, he might well have decided against giving the samples to Stacey.

The failures of these three organizations worked in synchrony to lead Stacey Palmer's family doctor to make the fateful decision to treat Stacey's headaches with Vioxx. But Mrs. Palmer's note to me raised a larger question about the system that had spawned this tragedy. Underlying Stacey's death was the sad truth that the very sources doctors now rely on to inform their clinical decisions have become profit-driven rather than science-driven, the facts manipulated to maximally benefit the pharmaceutical companies and their investors rather than the people taking their drugs.

Until the recent drug-company-fueled opioid crisis, Vioxx was the biggest drug debacle in U.S. history. The financial costs to Merck were significant. In 2007 the company agreed to pay $4.85 billion to settle civil litigation involving almost twenty-seven thousand plaintiffs who alleged Vioxx caused their heart attacks and strokes. And in 2011, the U.S. Department of Justice announced Merck had pleaded guilty to illegal marketing of Vioxx and paid almost $1 billion to settle criminal and civil allegations that it had marketed Vioxx off-label and made "false statements about the drug's cardiovascular safety."

So, Merck had been chastised. Surely the company changed course and became more forthright and trustworthy, correct? Not if you listen to Merck's current CEO, Ken Frazier, who served as Merck's general counsel during the years of Vioxx litigation. When asked by the *New York Times* in 2018 what lessons he had learned from Merck's defense of Vioxx, he responded: "The central allegation that was underneath the Vioxx litigation was that this company put profit before patient safety. That went to the core of who we are as a company, and there was no way that we could allow that story to be unchallenged. We did not want Vioxx to become a verbal shorthand, like an Enron or something of that nature, for corporate wrongdoing. The most important role of a leader is to safeguard the heritage and values of the company."

The "heritage and values" of a pharmaceutical company ought to be safeguarded by ensuring that doctors are given complete and accurate information about its products, so patients can receive full benefit without being exposed to unnecessary harm. Had Merck done that with Vioxx, Stacey Palmer would now almost certainly be in her thirties, and an estimated thirty thousand or more deaths in this country that resulted from taking Vioxx *after* Merck knew of the increased cardiovascular risk would not have occurred. Mr.

Frazier's decision to prioritize his company's reputation over fulfilling its responsibility to the doctors who prescribe Merck's products and the patients who take them was, as we will see in upcoming chapters, not a one-off. Rather, it was a sign of the times and a harbinger of the reckless pharmaceutical profiteering to come.

The industry's commercially motivated manipulation of clinical trial data comes into sharper focus in the next chapter, where trial testimony (including mine) will reveal the techniques of deception used by the largest drug company in the world as it violated federal racketeering laws.

2

NEURONTIN: FRAUD AND RACKETEERING

Some will rob you with a six-gun,
And some with a fountain pen.

— WOODY GUTHRIE, "Pretty Boy Floyd"

I n 2003, a tall, fit, conscientious medical student whom I'll call Steve presented a case to the primary care tutorial group I was teaching. The patient he discussed was being treated with Neurontin, a drug from Pfizer that had been approved by the FDA in 1994 as a secondary treatment for epilepsy and then in 2002 for persistent nerve pain after herpes zoster (shingles). Steve shared how taken aback he was when the neurologist he was working with said of Neurontin: "There is no other drug being used to treat so many different conditions with so little benefit."

To understand the neurologist's concern, we need to be aware that the use of Neurontin to treat these "many different conditions" — except for the two approved by the FDA — was off-label. Prescribing a drug off-label means using an FDA-approved drug to treat a condition that the drug has not been specifically approved to treat. In 2001, 21 percent of all prescriptions in the United States were off-label. And, as the neurologist suspected, the drug most frequently prescribed off-label that year was Neurontin.

Although off-label *prescribing* is not illegal, the *marketing* of drugs by manufacturers to treat off-label conditions *is*,* and with good reason. It turns out that almost three-quarters of off-label prescriptions written by U.S. physicians have "little or no scientific support." In other words, although journal articles (often sponsored by the manufacturer) may report that a given drug is efficacious for an unapproved use, until it has been formally reviewed and approved by the FDA for that use, the drug generally cannot be relied upon to provide effective and safe treatment for that condition. As this chapter will show, Pfizer got caught red-handed in an off-label-marketing scheme that accounted for the vast majority of its Neurontin sales. But the financial penalties for this scam, even when scientific and marketing malfeasance is proven in a court of law, are rarely higher than the profits made. And thus, this rapacious and sometimes deadly game of cat and mouse goes on.

KAISER FOUNDATION HEALTH PLAN V. PFIZER

Fast-forward seven years to 2010 and the U.S. District Court in Boston, where Pfizer, then the world's largest pharmaceutical company, was being sued by Kaiser Foundation Health Plan, the nation's largest HMO, for Pfizer's alleged off-label marketing of Neurontin. Serving as an expert witness for the plaintiffs, I stood next to the jury box and drew a graph to explain the statistical shenanigans Pfizer had used to persuade doctors to prescribe Neurontin off-label for nerve pain.

As mentioned, the drug had two FDA-approved indications, ep-

---

* With one exception: If a doctor requests information about a specific off-label use, drug companies may provide relevant articles.

ilepsy and nerve pain following shingles (post-herpetic neural-
gia). Kaiser alleged that Pfizer had pushed doctors to use Neurontin to treat bipolar disorder and to prescribe dosages up to twice
the FDA-approved maximum for *all* types of nerve pain, despite the
fact that the company's own science had failed to provide convincing evidence of benefit for these off-label uses.*

Kaiser's burden in proving the alleged wrongdoing was substantial: To win the case, lawyers had to prove not just that the doctors
didn't know the truth about Neurontin but that they *couldn't* have
known the truth about Neurontin due to Pfizer's manipulation of
the scientific evidence that the company alone controlled. Pfizer
had convinced doctors to prescribe Neurontin off-label for bipolar
disorder by delaying publication and ignoring the results of its own
study, which had shown the drug was *significantly worse* than placebo. For non-shingles-related nerve pain, Pfizer used other techniques, which included misrepresenting the results of one nerve-
pain study, rigging another, and suppressing the results of two
more. That they knew better was evident from an internal e-mail
from Pfizer's own medical director, which disparaged Neurontin as
"the 'snake oil' of the twentieth century."

Kaiser also claimed that Pfizer's deceptions had been perpe-
trated through a racketeering enterprise and violated the federal
RICO law (enacted in 1970 to curtail the activities of organized
crime). If the jury found Pfizer guilty of fraudulently influencing
physicians through participation in racketeering activity, the financial damages would be tripled.

In 2003, the year my student presented his case, annual sales of
Neurontin had reached $2.1 billion in the United States alone. And
Steve didn't know the half of it; as I noted in the report I submitted to
the court in advance of my testimony, in that year, nine out of ten prescriptions for Neurontin written in the United States were for non-

---

* Neurontin was originally marketed by Parke-Davis, a subsidiary of War-
ner-Lambert, which was acquired by Pfizer in 2000. For simplicity, I attri-
bute all research and marketing activities regarding Neurontin to Pfizer.

FDA-approved uses, a testament to the success of Pfizer's off-label marketing strategy. Further, owing to the unique effectiveness of aggressive marketing of new drugs in the United States, 86 percent of worldwide Neurontin sales in 2001 had been in this country.

For most of the year preceding the trial, I worked with a team of lawyers, combing through literally millions of pages of Pfizer corporate documents. These documents allowed me to compare the data from Pfizer's clinical trials — the real data — to what doctors were reading in medical journals. I also compared Pfizer's data to its internal marketing documents to determine whether Pfizer's "educational" meetings and conferences had presented doctors with a reasonably balanced version of the truth about Neurontin.

About a week before the trial began, I asked Tom Sobol, the Kaiser attorney who would question me in court, when we were going to prepare. I had worked with Tom on other drug cases, but I had never testified in an open trial with a live jury, and I expected to spend at least one intense day preparing with him. I was stunned when Tom, a fastidious lawyer, responded that we were not going to meet beforehand; we were just going to have a "conversation in the courtroom."

Although I was not scheduled to testify until the second day of the trial, I went to the federal courthouse the first day to hear the lawyers' opening statements. After passing through security, I noticed an inscription carved into stone on the atrium wall: THE RIGHT TO SUE AND DEFEND IN THE COURTS IS THE ALTERNATIVE OF FORCE. IN AN ORGANIZED SOCIETY, IT IS THE RIGHT CONSERVA- TIVE OF ALL OTHER RIGHTS, AND LIES AT THE FOUNDATION OF ORDERLY GOVERNMENT. This statement, made in 1907 by the U.S. Supreme Court justice William H. Moody, was a sobering reminder that the trial would be about far more than the fraudulent marketing of a single drug. Ultimately, it was about whether drug companies had the right to withhold and misrepresent crucial information about their products and thereby mislead doctors and the public.

Before the jury entered the courtroom, federal district judge Patti B. Saris laid down some ground rules. The testimony in this trial would remain focused on Neurontin, she said, not on the pharmaceutical industry in general. Therefore, no drug-company bashing would be allowed. Then she asked (in a stern voice), "Is Dr. Abramson in the courtroom?" I raised my hand from the gallery. She looked directly at me: "That means you."

I initially felt intimidated but then realized why she had singled me out. Two years earlier, I had made an unsolicited call to the U.S. Attorney's Office in Boston to see if they might be interested in the findings in my expert report on Bextra (another Pfizer product). I'd explained I was bound by a confidentiality agreement and could not share my report unless I was subpoenaed by the Department of Justice. I was duly subpoenaed, and, as mentioned in the introduction, I presented my testimony under oath to the DOJ and the FBI. Six months later, I learned that Pfizer had pleaded guilty to a felony and been assessed a record-breaking $1.195 billion fine. Now in Judge Saris's courtroom, I certainly got the message: I was not to discuss any of that — or make general statements about the pharmaceutical industry's behavior — in this trial.

My testimony began the next day. After being sworn in, I sat down in the witness stand and noticed that the jury box, which was on the other side of the expansive courtroom, was so far away that I could barely see the jurors' faces. As I answered questions about my background and credentials, I wondered how effectively I was going to be able to communicate with them. When I lecture, I typically feel a few butterflies before starting, especially with big audiences, but I generally get comfortable as soon as I begin my talk and can see how the audience is reacting. Here, that audience was so distant, I knew that would be difficult.

Marketing Neurontin to Treat Bipolar Disorder

The first issue I testified about was Pfizer's promotion of Neurontin for treating bipolar disorder. Pfizer's Pande study (named after the lead researcher) tested the effectiveness of the drug for severe bipolar disease in a double-blind randomized controlled trial (RCT). The study enrolled 117 manic patients already receiving but not responding adequately to drug therapy.

Dr. David Kessler, FDA commissioner from 1990 through 1997 and former cochair of President Biden's coronavirus task force, testified before I did. He explained to the jury that in a *randomized controlled trial* patients were randomly assigned to one of two study groups, one of which would receive the active treatment, in this case Neurontin, and the other a placebo. This ensured that the two groups were similar at the beginning of the study, so differences in outcomes between the groups would most likely be due to their different treatments. Furthermore, in the Pande study, bipolar patients were assigned to the two groups in a "double-blind" fashion, meaning that neither the subjects nor the researchers knew which patients were receiving the active drug and which were getting the placebo. The results were straightforward: At the end of the study, manic symptoms were *significantly worse* in the patients treated with Neurontin than in those treated with placebo. The clinical trial was completed in 1997, but, it should be noted, the results were not published as an independent study until 2000.

Three more RCTs found Neurontin to be no better than placebo for bipolar disorder. Nevertheless, between February 1996 and November 1999, the use of Neurontin for bipolar disorder increased fiftyfold, going from 8,000 to 402,000 prescriptions filled annually. What had caused this stunning increase when the study results had not only failed to provide evidence that the drug was helpful but, in one case, actually demonstrated that it was harmful?

My written report showed how this had happened: Pfizer had "educated" doctors about the unsubstantiated off-label use of Neurontin to treat bipolar disorder. In the spring of 1998 Pfizer sponsored fifty CME "Psychiatry Dinners" in expensive restaurants. (*CME* stands for "continuing medical education"; most states require physicians to take fifty hours of CME a year to maintain their medical licenses. CME programs are often sponsored by drug and device manufacturers.) All the lecturers at these dinners had prestigious academic positions. During the same period, Pfizer held sixteen ninety-minute psychiatry teleconferences and hired a company to provide thirty medical-education seminars, which were attended by more than eleven thousand doctors.

The information presented at these meetings included a recommendation to treat bipolar disorder with Neurontin. A physician with the title "Distinguished Senior Scientist" included the slide shown below at a Pfizer-sponsored program at the 1998 U.S. Psychiatric and Mental Health Congress; it was also shown to the jury as I testified about these CME seminars.

This slide reports the results of the Young study, which was completed in 1997. It shows that more than half the patients treated with Neurontin responded with marked or moderate improvement (20 percent and 33 percent, respectively). In truth, since the number of patients in this study was only fifteen, this translated to positive responses from three and five patients, respectively. Further, all the patients in this small study were treated with Neurontin (there was no control group — no placebo, no comparison drug, nothing), and both the subjects and the medical staff knew this. As I told the

---

Gabapentin for Bipolar Depression

• Add-on (no antidepressants)

• Dose: 300–2400 mg/day (N=15)

• Response
   –Marked (HAM-D > 50%)   in 20%
   –Moderate (HAM-D > 25–50%)   in 33%

Young et al. Biol Psychiatry, 1997; 42:851–853

Results of small open-label study of Neurontin (generic name gabapentin) for bipolar depression shown at educational meetings sponsored by Pfizer *This slide is discussed in my testimony in Kaiser v. Pfizer, February 23, 2010; data published in Robb Young et al., "Acute Treatment of Bipolar Depression with Gabapentin," Biological Psychiatry 42, no. 9 (November 1997).*

jury, the most damning thing about Pfizer presenting this slide at its "educational" meetings in 1998 was that the much larger randomized controlled double-blind Pande study had been completed a year earlier, but its results were not published for two more years. Therefore, until 2000, even the most conscientious doctors would not have been able to find the results of Pfizer's gold-standard study, which showed that Neurontin was worse than nothing.

I explained to the jury that when doctors attend lectures presented by experts, we assume the information is accurate: "Our job is not to go back to the medical library and look up the Young article and see what the study really was. So it's critically important [that] doctors get presented a fair and accurate and balanced representation of scientific evidence."

This was particularly true for Pfizer's off-label claims, which, by definition, addressed drug use that the FDA had not reviewed and approved. I reminded the jury, "The doctors are on their own when they're prescribing off-label."

On their own, that is, except for what they were told by Pfizer. During her turn in the witness chair, plaintiffs' expert and economist Meredith Rosenthal, professor at the Harvard School of Public Health, showed that Pfizer's promotional marketing of Neurontin for the treatment of bipolar disorder was strongly linked to an increase in Neurontin prescriptions for that use. As she explained to the jury, her analysis showed that 99.4 percent of these prescriptions were the result of Pfizer's "fraudulent marketing."

Marketing Neurontin to Treat Neuropathic Pain

Next came Pfizer's marketing of Neurontin for the treatment of nerve pain. To unravel Pfizer's commercial deception, I needed to compare the company's documents showing each study's design and data analysis plan with its presentation of the results.

Pfizer's analysis of an early RCT — the Gorson study, completed in 1997 — claimed Neurontin had the potential to reduce diabetic nerve pain. The Gorson results were presented at a large confer-ence and published in the journal *Neurology* with this conclusion: "[Neurontin] may be effective in the treatment of painful diabetic neuropathy." This must have looked like a slam dunk to health-care professionals — nerve pain in patients with diabetes was difficult to treat, but Neurontin might provide relief. Why wouldn't a doctor want to treat patients suffering from chronic pain with a potentially effective non-narcotic drug?

But that wasn't what the study results actually showed. The re-port Dr. Gorson faxed to Pfizer in 1997 presented a far less enthusias-tic conclusion: "[Neurontin] is probably no more effective than pla-cebo in the treatment of painful diabetic neuropathy." My task was to explain how this discouraging but scientifically accurate state-ment from the lead researcher had morphed into the Pfizer-friendly "Neurontin may be effective." Again, it wasn't complicated.

This time, I needed to carefully explain the meaning of *controlled* in the term *randomized controlled trial*. The purpose of an RCT is not to see whether people treated with the study drug (Neuron-tin, in this case) are better at the end of the study than at the be-ginning; that would be an *uncontrolled* study (like the Young study described above). The purpose of an RCT is to see if those treated with the study drug experience *significantly more* improvement than those in the group treated with placebo. For the Gorson study, comparison of the change in pain level between the two groups was important because the placebo effect came into play: A subjective end point, like pain, is far more susceptible to distortion than an objective end point, like blood-sugar levels or changes on an elec-trocardiogram. Also, people tend to volunteer for trials when their symptoms are at their worst, so with just the passage of time, pain is likely to return to its average symptom level without any treatment (called "regression to the mean").

For these reasons, the Gorson study had been appropriately de-signed to compare the *difference* in improvement between the Neu-rontin- and placebo-treated groups. But Pfizer's conclusion relied only on the change in the level of pain of patients *within* the Neurontin group, not on the difference *between* the Neurontin and placebo groups, and by doing this, the company was violating its own study design.

This was the crucial point I needed to get across. But without being able to see how the jury was reacting to my explanation, I couldn't tell if they were understanding me. I started to explain Pfizer's infraction against good science again, thinking I could make it clearer, but out of the corner of my eye, I saw Tom Sobol hold up a blue marker. "May I interrupt you, Dr. Abramson? Would it be helpful to draw a brief diagram to show this or no?" And at that point, I noticed an easel had been set up right next to the jury box.

"Yes," I said, trying not to respond too enthusiastically.

I decamped from the witness stand and walked across the courtroom, past the judge, past the Pfizer and Kaiser lawyers, to the position I described near the start of this chapter, just a few feet from the jury. From my new vantage point, the jurors' faces were no longer distant, undifferentiated blurs. There was a young gentleman in a camel-colored sport coat taking notes; a man in the back with his name sewn on his shirt; a young woman in a Red Sox jacket; and an elderly man who was sitting rapt. The Pfizer attorneys and the rest of the courtroom were behind me, out of sight and mostly out of mind.

Once I took that blue marker, I was no longer an expert witness for the plaintiffs—I was a teacher explaining to twelve motivated students with obviously different backgrounds how Pfizer had violated the rules of good science and how that scientific sleight of hand had led doctors to believe that Neurontin effectively treated the pain of diabetic neuropathy when the company's study had not shown that.

Standing at the easel, I could observe each juror to make sure he or she understood me as I explained how Pfizer used statistics in a deceptive way. First I sketched a graph of the results of the Gorson study, with a line descending from left to right to show how pain had steadily decreased over each of the six weeks of the study in the patients treated with Neurontin—just as Pfizer had claimed. Then

I drew a second line for the placebo group, also descending from left to right — almost mirroring the first line. The two lines, representing the average pain level in the two groups each week, were similar. Both groups had improved. For Pfizer to say that the Neurontin group saw a decrease in pain was correct, but it wasn't the comparison the study had been designed to make. What the study actually showed was that there was not a significant difference between the improvement in the Neurontin and placebo groups, and therefore using Neurontin to treat patients with painful diabetic neuropathy was unlikely to be helpful.

The study had been designed to investigate Neurontin's effectiveness in treating pain (that was the "primary outcome measure"), and the results showed exactly what Dr. Gorson had written in his original report: that Neurontin was "probably no more effective" than a sugar pill to treat pain. But in Pfizer's account of the study, there was no mention that a placebo group had shown similar results, so doctors could be tricked into believing the pain reduction in the people treated with Neurontin was due to the drug when it might just as well have been the placebo effect combined with the tincture of time.

At that point, the judge told Tom Sobol that it was time to finish up testimony for the day. I looked over at the jury and could see they understood how Pfizer had misrepresented the results of the Gorson study. More important, they now understood that scientific malfeasance was not out of bounds in the hardball world of pharmaceutical marketing.

Tom asked me one more question: If, rather than comparing changes in pain levels between the Neurontin group and the placebo group, one compared — as the manufacturer had — only Neurontin from the beginning to the end of the study, "have you essentially gotten rid of your control [group]?" Tom wanted to make absolutely sure the jury grasped this idea.

I responded, "You have . . . so you don't have a randomized controlled trial."

At which point the judge said, "Thank you, see you tomorrow."

Tom had set me up to finish the day by stating unambiguously that Pfizer had not followed its own study design.

The next morning my testimony picked up with another Pfizer-sponsored study of Neurontin for diabetic nerve pain. This was the Backonja study (again named for the lead researcher), by far Pfizer's most influential study of nerve pain, though neither the biggest nor the most rigorously designed. This study randomized 165 people with diabetic nerve pain to receive either Neurontin or placebo for eight weeks in what was ostensibly a double-blind RCT. The results, according to the 1998 article published in the *Journal of the American Medical Association*, showed that, compared to placebo, Neurontin significantly reduced daily pain scores. Pfizer hired a public relations firm to "educate both consumers and medical professionals" about the benefit of Neurontin, delivering its commercially advantageous message to Americans eighty-five million times through TV and radio, newspapers and magazines, even video clips shown to captive audiences of airline passengers.

But the devil was in the details of the study's design. It was a "forced titration" study, meaning the dose of Neurontin was increased over four weeks from 900 milligrams up to 3,600 milligrams per day — twice the FDA-approved maximum of 1,800 milligrams per day — and the dose was increased even if people experienced pain relief at a lower dose. Not surprisingly, by the end of the study, more than half of the Neurontin group had experienced side effects compared to only 15 percent in the placebo group.

The problem was more than simply the number of people who had experienced side effects. As I explained to the jury, developing symptoms like dizziness or sleepiness could have tipped off study participants that they were receiving Neurontin rather than a placebo. And this unblinding could have introduced bias because people experiencing side effects would have surmised they were probably receiving Neurontin and so would have expected to get relief from their pain. With this in mind, the authors of the *JAMA* article conducted further analyses, ostensibly to make sure side effects had not biased the study results. First, they removed the records of subjects who had reported dizziness (and so presumably knew they were receiving Neurontin, not a placebo) from the analysis of pain scores; they found the results still showed Neurontin superior to placebo. Then they repeated the exercise, removing the records of subjects who had experienced sleepiness; again, the results were the same. Based on these two analyses, they concluded that the greater frequency of side effects in people treated with Neurontin and the possible unblinding which that created "did not account for the overall efficacy seen in the trial."

I had a brainstorming session with a younger doctor and a lawyer, trying to figure out why this study had found Neurontin helpful for diabetic nerve pain when Pfizer's other studies had not. We pored over the language and analyses in the *JAMA* article until the light bulb finally went on: The authors had pulled a fast one by removing the study participants who had experienced each of the two most common side effects *separately*. They showed that removing the 24 percent of people who had developed dizziness when treated with twice the FDA-approved maximum dose of Neurontin did not alter the study's results. Ditto for removing the 23 percent of people who developed sleepiness. But they never checked to see if comparing those Neurontin-treated patients who developed *neither dizziness nor sleepiness* (and so presumably did not know if they were receiving the study drug) with the placebo-treated patients still showed that Neurontin provided significant pain relief.

Because we were in litigation, the lawyers could request the individual patient-level trial data from Pfizer. Nick Jewell, professor of biostatistics and statistics at the University of California, Berkeley, and also a plaintiffs' expert in this trial, reanalyzed the results using the pain levels of participants recorded at the last visit *before* they experienced side effects. Using the pre-side-effect pain scores provided a statistical way to remove the bias created by forcing the dose of Neurontin up to twice the FDA-approved maximum and causing side effects that could have tipped off half the people treated with Neurontin that they had been assigned to the active-treatment group. Indeed, Professor Jewell's reanalysis showed that 90 percent of the improvement in pain attributed to Neurontin had occurred *after* the onset of side effects. Analyzing only pre-side-effect pain scores showed that the pain reduction associated with Neurontin was no better than placebo.

More than ten years after that highly publicized but misleading *JAMA* article was published, Professor Jewell and his colleagues published their findings in two biostatistics journals. Sadly, few if any doctors would be influenced by these highly technical statistical articles appearing many years after they had already accepted the results published in their trusted *JAMA*.

An important contrast is provided by another Pfizer study of Neurontin for diabetic nerve pain, the Reckless study (again named for the lead researcher), which had no trickery in its design. In this study, completed in 1999, three times more people were treated with Neurontin than in the Backonja study, and, instead of forced titration, this design was much hardier: Three groups treated with fixed doses of Neurontin — 600, 1,200, and 2,400 milligrams per day — were compared to patients treated with placebo. Pfizer's research report stated unequivocally that "there was no statistically significant difference between any of the gabapentin groups and the placebo group for end point mean pain score or at any time throughout the trial."

Published in a timely and forthright fashion, these results would have had a definite impact on doctors' beliefs about the efficacy of Neurontin for diabetic neuropathy. But in stark contrast to the Pfizer-sponsored PR blitz that followed publication of the Backonja study, the results of this far more important study made an impression on few consumers or medical professionals. This is because the results were never published as an independent article, robbing doctors of the opportunity to integrate these negative findings into their treatment plans. Pfizer did, however, issue an internal communication about the Reckless study, which I read to the jury. It stated: "Although I would love to publish

SOMETHING about [the study], Donna McVey [a Pfizer medical director] made it very clear that we should take care not to publish anything that damages neurontin's [*sic*] marketing success." Pfizer's Neurontin Publication Subcommittee agreed that the results of the Reckless study "should not be pushed for publication." Clearly, Pfizer's commitment was not to patients' welfare but to selling Neurontin, even if it meant withholding from doctors the most important clinical trial data about Neurontin's benefits (or lack thereof). Pfizer never did publish the results of the fixed-dose study as an independent article.

Despite the negative results of this well-designed study, Pfizer continued to pursue the holy grail of pain-medicine sales: FDA approval of Neurontin for the treatment of all types of nerve pain. But when they met with the FDA in May 2001 to discuss their application to make treatment of nerve pain an approved indication, the FDA — knowing what Pfizer's data showed — wouldn't even allow Pfizer to file the application.

Pfizer then did the right thing. In September 2001 its executives convened a meeting of independent pain consultants and requested honest advice about how to win FDA approval. After reviewing Pfizer's clinical trial data, both published and unpublished, the consultants were as disparaging as the FDA. One said there was "substantial evidence against a broad neuropathic pain claim." Another concluded concisely, "You're done." Ultimately, Neurontin never received FDA approval for the treatment of any type of pain other than persistent post-shingles pain.

THE MISUSE OF KEY MESSAGES

Meanwhile, Pfizer was pursuing another strategy, one that ignored the FDA and the company's own pain consultants and actively misled doctors about the effectiveness of Neurontin for neuropathic pain. This involved not just withholding negative clinical trial results but making affirmative claims about the drug that its own clinical trials had shown to be untrue. My access to the documents in this litigation provided a rare opportunity to see how blatantly this had been done.

In mid-July 2001, Pfizer started working with the consulting firm Medical Action Communications on developing key messages to be incorporated into a review article recommending that doctors prescribe Neurontin to treat neuropathic pain. First on the list of "Neurontin Publication Plan Key Messages" presented in an e-mail dated July 30, 2001, was "proven efficacy for neuropathic pain," which, as we've just seen, Pfizer was well aware had not been proven. The following day, another key message recommended increasing the dose of Neurontin to 1,800 milligrams per day—the FDA-approved maximum dose—by the second week of therapy *even if the patient was experiencing relief at a lower dose.* A third key message went even further: "Gabapentin doses up to 3600 mg/d have been proven well tolerated and effective in clinical studies." Just to be clear, this was two months after the FDA said that, based on the available evidence, it would not even consider an application for approval of Neurontin to treat neuropathic pain, two months before Pfizer's own pain consultants opined that the evidence did not support approval of Neurontin for the treatment of neuropathic pain, and in spite of 1,800 milligrams being the maximum daily dose approved by the FDA.

In January 2003 the review article was published in the journal *Clinical Therapeutics,* and its conclusion precisely reflected the key messages developed back in July 2001: "At doses of 1800 to 3600 mg/d, gabapentin was effective and well tolerated in the treatment of adults with neuropathic pain." A Pfizer e-mail told its drug reps: "Because this is a key publication for Neurontin," the information in that review article should be included in all marketing activities related to the treatment of neuropathic pain.

After I read this to the jury, a juror raised her hand, was recognized by the judge, and asked, "Is it legal to promote off-label applications? Is any of this legal?"

I responded:

> It is not legal to promote — to market off-label. . . . If a drug rep came in and said, "Hey, Neurontin is a good drug for adjunctive therapy for seizures," and the doctor asked, "Well, it's an anti-seizure medicine, might it work for neuropathic pain?" then the drug rep is allowed, having received an unsolicited request, to show information that would make that case. But unless the drug rep is specifically asked that question, it's not legal.

The juror responded, "So this was arming [the drug reps] with the information should they be asked?"

I answered, "No, I don't think that's true." The juror acknowledged my comment, and I added, "I think this was proactive," by which I meant the purpose of training the reps to tout the merits of treating neuropathic pain with Neurontin was not simply so they would be prepared to respond to an unsolicited question but to offer the "education" even *before* having received a specific inquiry.

The juror said, "Understood."

At the end of my testimony about neuropathic pain, Tom Sobol circled back to ask me why it was so important that doctors understand the bias that might have been introduced into the 1998 Backonja study of Neurontin for diabetic neuropathy by unblinding. Why, he asked, wasn't this "just sort of a geeky statistical" point?

Doing my best not to sound geeky, I explained,

> Well, that is really important because doctors work very hard, and they want to get the bottom line of the research: Does this study show that the drug works for patients with diabetic neuropathy or not? And these fine details . . . really change the meaning of the study. So it's essential for doctors to understand the fine print here, and yet . . . there are a few clues, but there's no way you can expect a practicing physician to unravel the incomplete correc-

tion for the unblinding that happened [because of] the forced titration design of the study.

I don't think I succeeded in not sounding geeky, but the jury
seemed to understand what I was saying.

THE VERDICT AND BEYOND

The trial lasted six weeks. After deliberating for two days, the jury
found that Pfizer had fraudulently promoted Neurontin to doctors
for off-label use, and it awarded Kaiser $47 million. The jury also
found — for the first time in a drug company case — that Pfizer had
violated the RICO Act (that is, it had committed racketeering violations), which automatically tripled the penalty to $142 million.
Pfizer appealed the decision and lost.

Pfizer hadn't acted like the gangsters of the past — no machine
guns, no bank heists, no hit men. Rather, as Judge Saris pronounced
in her "Findings of Fact and Conclusions of Law," Pfizer had engaged in a "nationwide effort to unlawfully market this drug for
off-label uses for which there was little or no scientific evidence of
efficacy."

The outcome of this litigation might look like a resounding de-
feat for Pfizer and a victory for the integrity of the data that doctors
rely on to make their clinical decisions. But time has shown just the
opposite. The financial penalties Pfizer paid in this and all the other
Neurontin litigation amounted to a relative pittance, less than half
the revenues from one year of Neurontin sales. No one went to jail,
and the press coverage of this trial was minimal, so Pfizer suffered
little reputational damage.

I had hoped that once the truth was presented in court, doctors
would understand how their patient care was being undermined by
Pfizer's illegal marketing of Neurontin. That didn't happen; among
the health-care professionals I speak to, only a small percentage
are aware this trial ever took place. A 2019 update of clinical tri-

als of gabapentin shows the clinical evidence hasn't changed significantly in the intervening years and concludes that "clinicians who prescribe [gabapentin] off-label for pain should be aware of the limited evidence and should acknowledge to patients that potential benefits are uncertain for most off-label uses." Nonetheless, once doctors get in the habit of prescribing a certain drug to their patients, their belief in its purported benefit takes on an indelible quality — even when the source of that belief is the manufacturer's illegal marketing of the drug. The result: Even today, generic Neurontin (gabapentin) is the sixth most frequently prescribed drug in the United States, and most of those prescriptions are for off-label use.*

The *Kaiser v. Pfizer* trial provided two key lessons about drug company marketing: First, because drug companies fund most of the research about their own drugs and control the resulting data, they can (and do) mislead physicians in order to increase sales. And second, under our current system, it is more profitable for large pharmaceutical companies to commit crimes and pay the fines than to obey the law. Why dedicated doctors can be so predictably misled will be addressed in part II.

---

* Some of the overreliance on gabapentin to treat pain is a reaction to the recent overprescribing of opioids.

3

THE TRUTH ABOUT STATINS

The statins saga forces us to confront the deep flaws in our current system for evaluating medicines and guiding clinical decisions. In particular, how can it be right to recommend mass treatment of healthy people without independent review of the patient level data?

— EMMA PARISH, THEODORA BLOOM, AND FIONA GODLEE,
British Medical Journal

The first two chapters showed how sales of individual drugs were jacked up by their manufacturers' manipulation of the scientific information made available to (or withheld from) doctors. This chapter presents a similar phenomenon but for an entire class of drugs: cholesterol-lowering statins, which are by far the most frequently prescribed class of drugs in the United States. As with most widely used classes of drugs, doctors prescribe statins based on clinical practice guidelines issued by their professional societies and relevant nonprofit organizations.

To understand the importance of those guidelines, meet Jane, a bright, hardworking, and socially committed baby boomer. Jane married in 1970, soon after graduating from college, and spent the next four decades doing it all; she was a wife, the mother of three, and a dedicated full-time fifth-grade teacher for twenty years, first in a rural town in the Pacific Northwest and then in a suburb of Boston. Every weekday, teaching and family responsibilities filled her waking hours; after dinner, she would return to teaching chores until around 9 p.m. Despite being aware that the unrelenting demands on her time and energy were creating stress, she acknowledged that "taking care of myself was the last thing on my list of things to do."

When Jane reached her early fifties, her primary care doctor became concerned that her elevated cholesterol level was increasing her risk of developing heart disease, so she spoke with Jane about the importance of eating a healthy diet, exercising, and cutting out the few cigarettes she was smoking each day. A couple of years later, in 2002, dutifully following the guidelines issued the previous year, Jane's doctor started her on Pravachol (pravastatin), a moderate-strength cholesterol-lowering statin. And two years after that, when Jane's cholesterol still had not come down to target levels, her doctor switched her to a more potent statin, Lipitor (atorvastatin), beginning with the lowest dose, 10 milligrams a day, and gradually raising the dose to the maximum, 80 milligrams a day.

At Jane's annual physical exam in August 2008, her doctor again recommended that she lose some weight and give up the two cigarettes a day she was still smoking. She also reassured Jane that her cholesterol level and blood pressure were both excellent. That's why Jane wasn't concerned when, just five days later, she developed what felt like mild indigestion while driving home from a visit with her brother's family, who lived several hours away. She recalled having the sensation that "if I could just belch, I would feel better." By the time she pulled into a rest stop an hour later, Jane's "indigestion" had progressed to pain in her jaw and beneath her breastbone. A nurse who happened to be in the ladies' room with Jane noticed she was sweating and nauseated and recommended she get checked out as quickly as possible. Jane's husband took over the driving, and they raced to their local hospital.

In the emergency department, physicians determined that Jane was having a major heart attack. She was immediately transferred to a Boston teaching hospital, where a cardiac catheterization revealed an acutely blocked left anterior descending coronary artery (often called the "widow maker" because blockage of this vessel causes the majority of fatal heart attacks). Fortunately, an interventional cardiologist was able to open the artery with angioplasty and insert a stent to keep it open.

Jane recovered well, worked hard in the cardiac rehabilitation program at her local hospital, and began exercising daily at the local YMCA. But the damage to her heart left her vulnerable to further complications. Seven months after the heart attack, Jane was washing dishes following a family dinner with her two grandchildren when she suddenly began seeing silver and gold sparkles in front of her eyes. She described feeling "like I was losing power from the top of my head down." Her husband called 911. The paramedics arrived in two minutes, put her on a stretcher, and hooked her up to a cardiac monitor. It showed ventricular tachycardia, a life-threatening arrhythmia in which the heartbeat originates in a ventricle rather than coming down from the atria through the normal conducting fibers. They raced her out to the ambulance parked in front of her house. She was still conscious when the paramedics delivered a shock to convert her heart back to a normal rhythm. "The electricity hit me in the head, powerful and bright." But it didn't work and left her with an even more dangerous heart rhythm. She lost consciousness. The second shock was successful, returning her heart to a regular rhythm and saving her life.

In the hospital, Jane received an implantable defibrillator: Electrical lead wires were inserted through a vein into her heart and then connected to a stopwatch-size device that was placed under the skin of her upper chest. The device would monitor her heart rhythm and automatically deliver a shock if a dangerous arrhythmia occurred. After a couple of days of in-patient cardiac monitoring, Jane was discharged. Fortunately she has had no further heart problems. She continues to exercise about five times a week, is maintaining her forty-pound weight loss by sticking to a healthy diet, has finally quit smoking, and feels better than she has in decades. Her heart rhythm has remained normal, and so far the services of the implanted defibrillator have not been needed. And now, in a higher-risk category because she's *had* a heart attack, she continues on a statin for what's called "secondary prevention."

Did Jane's heart attack fall into the category of "bad things can still happen despite good care"? Or did she have a heart attack because, unbeknownst to her doctor, there was a gap between the recommendations in the clinical guidelines and what was actually the most effective care?

THE GUIDELINES WERE PART
OF THE PROBLEM

In 2002, when Jane was started on pravastatin, her doctor had good reason to believe that a statin would, by lowering Jane's cholesterol level, lower her risk of heart disease. That was the conclusion of the highest authority: the evidence-based guidelines issued in May 2001 by the National Cholesterol Education Program (NCEP) coordinated by the National Heart, Lung, and Blood Institute (NHLBI). The guidelines recommended statin therapy for healthy men and women who, like Jane, did *not* have a history of heart disease or stroke but *did* have two or more major risk factors,* a ten-year risk of heart disease between 10 and 20 percent, and an LDL level of 130 mg/dL or higher despite attempts at lifestyle modification.

Compared to the earlier guidelines from 1993, these criteria called for almost three times as many Americans to be taking cholesterol-lowering statin therapy. And most of the people for whom statins were newly recommended were people, like Jane, who were healthy but did have some evidence of increased risk.

The twelve-page executive summary of these guidelines, pub-

---

* Major risk factors for heart disease were identified as cigarette smoking, high blood pressure, low "good" cholesterol (HDL less than 40 mg/dL), family history of early heart disease, and age (men forty-five or older, women fifty-five or older).

lished in the *Journal of the American Medical Association*, assured doctors that clinical trials of statin therapy had shown "favorable effects" in women with or without a previous history of heart disease. But buried on page 211 of the full-length version of these guidelines was contradictory evidence that few doctors would ever see. A table titled "Special Considerations for Cholesterol Management in Women (Ages 45–75 years)" reported that for women at Jane's risk level, "Clinical trials of LDL lowering *generally are lacking for this risk category;* rationale for therapy is based on extrapolation of benefit from men of similar risk" (italics mine).

In plain English, this means that clinical trials had *not* shown that statins had "favorable effects" on women; in fact, the experts admitted there had not been enough women in the original clinical trials to determine what, if any, effect statins had on women. The authors of the guidelines had simply decided that, if the effects were favorable for men ages forty-five to seventy-five, they would also be favorable for women. This was unacceptable. It was 2001, and statins had been on the market for fourteen years. If the drug companies wanted to claim their statins were beneficial for the additional millions of healthy women included in the new guidelines, the burden was on them to do the clinical trials to prove it. Simply assuming that the benefits of cholesterol-lowering statin therapy for men in that age category applied to women was not scientifically valid. Nevertheless, this is exactly what the guideline authors had done.

In 2004 the guidelines were updated again, based on the results of five additional trials that had become available since 2001. The update recommended that for women at Jane's level of risk "an LDL-C goal [of less than] 100 mg/dL is a therapeutic option on the basis of recent trial evidence"; previously, the threshold had been 130 mg/dL. Jane's doctor, wanting to provide her with the best care possible, dutifully increased the dose of Lipitor to bring her LDL level down to below 100 mg/dL.

But closer inspection of the five studies that constituted the "recent trial evidence" revealed that only one had addressed the benefit of statin therapy in women, like Jane, with multiple risk factors and no history of heart disease. In the Anglo-Scandinavian Cardiac Outcomes Trial (ASCOT), published in 2003, men and women with hypertension and three or more additional risk factors for heart disease were randomly assigned to take either 10 milligrams per day of Lipitor (atorvastatin) or a placebo. Lipitor failed to reduce the risk of death in men and women combined, and the women treated with Lipitor developed *10 percent more heart attacks* than the women treated with placebo. This difference did not reach statistical significance, but it certainly did not provide evidence that statins were beneficial for healthy women like Jane (prior to her heart attack), nor did it provide evidence to support the lower LDL target for healthy women.

When I found that eight of the nine panel members who had written the 2004 guidelines update had financial ties to statin manufacturers, I contacted Merrill Goozner (author of *The $800 Million Pill*), and together we wrote a letter to the National Institutes of Health that was signed by thirty-five physicians and researchers. We requested reconsideration of, among other issues in the update, lowering the LDL target for women without heart disease. We also expressed concern about the undisclosed conflicts of interest among the guideline authors.

Dr. Barbara Alving, acting director of the NHLBI, responded to our letter, acknowledging "the absence of specific clinical trial evidence" to justify the recommendation to treat healthy women with statins but lamenting that "for many women, the first sign of heart disease is sudden death." Therefore, wrote Dr. Alving, "sound public health policy demands that the significant risk for illness and death in women be addressed with science-based prevention recommendations." We agreed wholeheartedly, but our point was there was *no* science-based evidence from clinical trials showing that statins reduced healthy women's risk of dying of heart disease. And there still isn't. The NHLBI has never acknowledged this flaw in its logic.

Dr. Alving did, however, acknowledge the "desirability of having financial disclosure information as publicly available as possible," and the financial ties the authors had with statin makers were posted on the NHLBI website.

In 2007 Dr. Jim Wright, director of the University of British Co-lumbia's Therapeutics Initiative,* and I published an article in *The Lancet* titled "Are Lipid-Lowering Guidelines Evidence-Based?" We wrote that the seven studies listed in the 2001 NCEP guide-lines claiming to provide evidence that statins were beneficial for healthy women did nothing of the sort. Six of the seven studies in-cluded only women who already had heart disease (and were thus in the high-risk, secondary-prevention category), and the seventh study showed statins reduced neither the risk of heart attack in women nor the overall risk of death.

I expected the statin world to stop spinning on its axis, but there was barely a peep in response to our article.

2013: STATINS FOR EVEN MORE HEALTHY PEOPLE

The recommendations in the 2001 and 2004 cholesterol guidelines that led Jane's doctor to prescribe a statin for her and then increase the dose of that statin remained unchanged for nearly a decade. Then, in November 2013, the American College of Cardiology and the American Heart Association (ACC/AHA) issued new choles-terol guidelines. The recommendations for secondary prevention and for people at high risk of cardiovascular disease didn't change much, but once again, the threshold for treating healthy people with statins was lowered; statins were now recommended for all people with 7.5 percent or greater ten-year risk of cardiovascular disease (replacing the previous threshold of 10 percent or greater), and the requirement that LDL had to be above a certain level was removed. These changes increased the number of healthy Ameri-

---

* The UBC's Therapeutics Initiative is one of the best open-access sources of unbiased drug information; see http://www.ti.ubc.ca/.

cans for whom statin therapy was recommended by more than 70 percent and now included almost a quarter* of all healthy Americans ages forty to fifty-nine and at least two-thirds of those who were sixty to seventy-five.

In response to the new guidelines, Rita Redberg, professor of cardiology at the University of California, San Francisco, and editor of *JAMA Internal Medicine,* and I wrote an op-ed for the *New York Times* titled "Don't Give More Patients Statins." We argued that, for the low-risk population (less than 20 percent ten-year risk of a cardiovascular event), the evidence did not show that statins reduced the risk of death or serious illness, and the reduction in nonfatal cardiovascular events was small. We concluded: "Instead of converting millions of people into statin customers, we should be focusing on the real factors that undeniably reduce the risk of heart disease: healthy diets, exercise and avoiding smoking."

Our concern was that increasing the number of healthy people for whom statins were recommended would distract even more patients from making the lifestyle changes that were far more likely to reduce their risk of heart attack and strokes and distract even more doctors from helping their patients make those healthy lifestyle changes. Which raises the question: Are these guidelines really evidence-based? By the widely accepted definition of *evidence-based medicine* — reliance on peer-reviewed reports of methodologically high-quality clinical trials and systematic reviews published in medical journals — the answer would appear to be yes. As described by Dr. Jennifer Robinson, vice-chair of the 2013 guidelines panel: "We relied extensively on the Cholesterol Treatment Trialists [CTT] meta-analyses . . . in making our treatment recommendations."

But if we are asking whether there is fully transparent, independently analyzed evidence supporting these updated recommendations, the answer is quite different. Let's take a closer look at those CTT meta-analyses. (Meta-analyses combine the data from

---

* The numbers increase if the optional 5 percent ten-year-risk threshold is used.

relevant studies.) The CTT Collaboration was established back in 1994 for the purpose of pooling the data from all the completed major statin trials periodically into meta-analyses. As stated in the CTT's protocol, issued the next year, the raison d'être of the meta-analyses (there would be several over the coming years) was that, although no single trial would be large enough (that is, have the "statistical power") to determine the effect of statin therapy on overall deaths, the "unequivocal evidence about the net effects of several years of treatment on total mortality" could be achieved by combining the results "of all current and planned randomized trials." Additionally, the CTT expected the pooled data to provide increasingly reliable evidence of the effect of statin therapy on the risk of heart disease in "special interest groups," such as those at higher or lower risk, women, and the elderly.

In the same way that Pfizer kept the underlying data from its Neurontin studies strictly confidential, the statin makers have kept the results from their clinical trials confidential. Further, by submitting their underlying data to the CTT Collaboration (with the agreement, as stated in its protocol, that the collaboration would not release the data to anyone else), statin makers could have their cake and eat it too. The increased statistical power of the pooled data would likely provide evidence that would increase sales of all statins without giving an advantage to any single brand. The rising tide would lift all boats, so all the manufacturers would benefit. Since 2005 the CTT Collaboration has published five statin meta-analyses in *The Lancet,* each one increasing the population that would purportedly benefit from taking a statin.

But the problem — and it is a foundational problem — is that few doctors understand that the underlying data from the clinical trials and meta-analyses, relied on in the 2013 guidelines, *were not available to the experts who wrote those guidelines.* This fact cannot be emphasized too strongly. And I can testify to doctors' lack of awareness of this fact because I conducted an informal survey of fifteen hundred health-care professionals while delivering a keynote address to the annual meeting of directors of long-term care facilities.

I asked for a show of hands: How many people were aware that the experts who wrote the 2013 cholesterol-lowering guidelines recommending statin therapy for tens of millions of healthy Americans had not been given access to the actual data from the clinical trials? A single doctor raised his hand. From the podium I asked how he knew. He said he had read in my book *Overdo$ed America* that the authors of the 2001 guidelines did not have access to the underlying data. I didn't want to embarrass him in front of his colleagues, but he was wrong. It was not in my book. I had studied those guidelines very carefully, but in 2004, when the book was published, I had not yet discovered that the guideline writers did not have access to the underlying data. Nor did I understand, until I served as an expert in litigation, how necessary unfettered access to the underlying data from clinical trials is to enabling verification of the accuracy and completeness of published reports.

So why, I hope you are wondering, would the fifteen experts who wrote the 2013 cholesterol-lowering guidelines be willing to recommend medication for half of all Americans between ages forty and seventy-five without having had access to the clinical trial data?

Perhaps the answer has something to do with the fact that eight of the fifteen members of the expert panel had financial ties to drug companies at the time they were asked to participate. Of these eight, one was the chair, Professor Neil Stone, who severed ties to six drug companies (all of which manufactured cholesterol-lowering medications) after being asked to lead the panel. Another was one of the two vice-chairs, Dr. Jennifer Robinson, who declared financial relationships with at least nine drugmakers during the time the guidelines were being written. In the three years following release of the 2013 ACC/AHA guidelines, Dr. Robinson received more than $3 million from the drug industry for consulting and research.

These conflicts of interest were present despite the Institute of Medicine's 2009 recommendation "that chairs of guideline development panels" — vice-chairs were added in 2011 — "have no conflicts of interest, limiting members with conflicts of interest to a small minority of the panel membership, and precluding such members from voting on topics in which they have a financial interest." In addition, the AHA and the ACC, the two sponsoring organizations of the guidelines, both rely on industry support.

STATINS FOR PEOPLE AT HIGH RISK

There is no question that statin therapy can be of significant benefit to people at the highest risk of a cardiovascular event — that is, those who have already had a heart attack or a stroke or who have other evidence of arterial blockage. The CTT meta-analysis of 2012 reported that people treated with statin therapy for secondary prevention had 20 percent less risk of having a heart attack or stroke over five years of therapy than those treated with placebo. Similarly, people who were taking a statin for secondary prevention had 12 percent less risk of dying than those taking a placebo.

But these percentages describe *relative* risk reduction and do not provide the information patients and doctors need to engage in rational shared decision-making about the chance that any *one* individual will benefit from taking a statin. For that we need to consider the *absolute* risk reduction — that is, the difference between the risk of a second heart attack or stroke in people who don't take a statin (17.6 percent over five years) minus the risk for people who do take one (14.3 percent over five years). So a person with a history of cardiovascular disease who takes a statin for five years reduces his or her personal risk of suffering a recurrent episode by 3.3 percent, meaning that the absolute risk reduction was 3.3 percent.*

---

* Heart attacks and strokes are considered "hard" cardiovascular outcomes, meaning that these diagnoses are typically neither ambiguous nor vulnerable to subjective variation. "Soft" cardiovascular events, meaning cardiac revascularization procedures, and the CTT's post hoc composite end point of "major vascular events" that incorporates them, are not considered in these calculations for two reasons. First, soft outcomes are usually neither as serious nor as diagnostically objective as the hard outcomes of heart attack and stroke. This is in part because lower LDL levels in the people being treated

A more practical way for doctors and patients to discuss risk reduction is by using the "number needed to treat" (NNT), meaning, in this case, the number of people who must be treated with a statin to prevent one heart attack or stroke. NNT is easily calculated by dividing 100 by the absolute risk reduction (3.3 percent); the result, thirty, indicates that thirty secondary-prevention people must be treated for five years in order to prevent one heart attack or stroke; the other twenty-nine people will derive no benefit. Unfortunately, there is no way of knowing in advance which one of the thirty people will benefit; however, all thirty will have been exposed to the risk of side effects of the drug (discussed below). Similarly, the absolute risk reduction of death among people treated with a statin who already have cardiovascular disease is 1.25 percent over five years, so to prevent a single death with five years of statin therapy the NNT is eighty (NNT = 100/1.25). Again, statin therapy will not protect the other seventy-nine from death.

Most doctors think a one-out-of-thirty chance of preventing a serious cardiovascular event and a one-out-of-eighty chance of preventing a death merit recommending treatment with a statin for people with a history of cardiovascular disease. I agree. Still, I believe that these numbers should be presented to secondary-prevention patients to allow them to make informed choices. The decision about continuing to take a statin becomes more nuanced when a secondary-prevention patient is experiencing side effects. As with any discussion about reducing the future risk of heart disease and stroke, doctors should make sure their patients understand that the benefits of healthy lifestyle changes are likely to be far greater than

---

with a statin in clinical trials partially unblind diagnostic and treatment decisions. Second, the composite end point of "major vascular events" was not prespecified in the protocol for the CTT meta-analyses. Later addition of this composite end point was prohibited as an outcome measure because it had not been specified prior to the unblinding of the data, which allowed the CTT researchers to know the results for this outcome before they committed to reporting it.

the benefits of cholesterol-lowering medication, though the two are not mutually exclusive.

STATINS FOR PEOPLE AT LOW RISK

Determination of the benefit of statin therapy for people at low risk of cardiovascular disease, meaning people like Jane before her heart attack, is more complicated. For one thing, clinical trials of statins for people at low risk generally limit enrollment to people with no history of cardiovascular disease (so-called primary prevention). But many of the people in these studies do not really qualify as "low risk" because — though they may not have a history of cardiovascular disease — they do have greater than a 20 percent risk of suffering a cardiovascular event over the next ten years. For example, more than half of the cardiovascular events in two major primary-prevention trials — the aforementioned ASCOT study and the West of Scotland Coronary Prevention Study (WOSCOPS) — occurred in people with greater than 20 percent ten-year risk. The guidelines already recommended statins for people above this level of risk, and mixing them in with a lower-risk population artificially inflates the benefit of statins in low-risk people.

In 2012, the CTT published a meta-analysis designed specifically to determine the net effect of cholesterol-lowering "with statin therapy in people at low risk of vascular disease." (Remember, only the CTT had access to patient-level data from most clinical trials, so only the CTT could calculate the effect of statin therapy among people whose ten-year risk was less than 20 percent.) But the CTT meta-analysis did a bait and switch, concluding statin therapy reduced the risk of death for people *who did not already have cardiovascular disease*. They simply failed to address the question the article had set out to answer: Do statins reduce the risk of death for people at low risk (ten-year risk of less than 20 percent) of cardiovascular disease?

Because the findings from the 2012 meta-analysis were sure to play a pivotal role in determining who should receive statin therapy — since the 2013 ACC/AHA cholesterol-lowering guidelines were soon to be issued — Professors Jim Wright, Nick Jewell, Harriet Rosenberg, and I decided to recalculate the CTT's 2012 findings specifically to answer the question that the CTT had sidestepped: Does statin therapy provide a net benefit for people at low risk of cardiovascular disease? To do this, we recalculated the effect of statin therapy on overall mortality and on heart attacks and strokes in people whose risk of cardiovascular disease was less than 20 percent over the next ten years.

Our findings, published in the *British Medical Journal* in October 2013, showed that statin therapy provides very little or no net benefit in this population:

- no significant reduction in mortality (the overall risk of death)
- small (though statistically significant) reduction in the risk of nonfatal heart attack and stroke — 140 people with low risk (less than 20 percent five-year risk) must take a statin for five years to prevent one nonfatal event (NNT = 140)
- no reduction in serious adverse events (events serious enough to cause hospitalization) overall

These findings received extraordinary scrutiny after Professor Sir Rory Collins, head of the CTT Collaboration, demanded retraction of our article.* Dr. Fiona Godlee, editor of the *British Medical Journal,* appointed an external panel of six experts to adjudicate Professor Collins's demand. Although we trusted our primary findings were correct, the process, which lasted from May through August 2014, was harrowing. We were up against the most powerful pro-statin advocates in the world. When the *BMJ*'s verdict was fi-

---

* Professor Collins's call for retraction was based on an unrelated error in calculation of the risk of statin-related side effects. This error had no impact on the primary finding of the paper, which was that statins offer minimal benefit for healthy people.

nally delivered, their two independent statistical opinions had confirmed our calculations, and "the panel was unanimous in its decision that the [article did] not meet any of the criteria for retraction."

More important, our re-analysis of the CTT data had clearly shown that the net benefit of statin therapy for low-risk people was minimal. Approximately one hundred low-risk people (between 7.5 percent and 20 percent five-year risk) must be treated with a statin for five years to avoid one nonfatal heart attack or stroke without reduction in the risk of serious illness or death. In other words, 99 percent of low-risk people who qualify for statin therapy under the 2013 guidelines and who take a statin for five years will derive no benefit from it.

Again in June 2020, the same three coauthors of the 2013 *BMJ* article and I responded to CTT members' claim, based on their 2019 meta-analysis, that all people over age seventy-five should be taking a statin, whether or not they have a history of cardiovascular disease. Our letter, published in *The Lancet*, again based on review of CTT data, stated that "annually, 1000 people older than 75 years without a history of vascular disease need treatment to prevent a single major vascular event, and cardiovascular or all-cause mortality data are not presented for this population." CTT members responded, but they did not dispute these calculations. This means that if a thousand people over the age of seventy-five without a history of cardiovascular disease are treated with a statin for one year, *999 will derive no demonstrated benefit* and will be exposed to the risk of side effects. This is how distorted our health care has become: Despite the absence of data showing that statin treatment in elderly patients without heart disease provides significant benefit, the percentage of Americans without cardiovascular disease age seventy-nine and older taking a statin more than tripled between 2000 and 2012, from 8.8 to 34.1 percent.

For those who are not at high risk of cardiovascular disease, regardless of age, the real harm in taking a statin is not the expense or even the exposure to potential side effects. The real harm is the false reassurance of protection from cardiovascular disease the guidelines provide, distracting both doctors and patients from the harder but more effective work of adopting healthy lifestyle habits. Like so many others, Jane believed she was "off the hook" after starting statin therapy, and she gained weight and exercised less.

According to the World Health Organization, 80 percent of premature heart disease can be prevented by adopting just three habits: eating a healthy diet, exercising regularly, and avoiding tobacco. Similarly, the U.S. Nurses' Health Study* suggests that more than 80 percent of heart disease in women can be prevented if they adopt the same three healthy habits, along with moderating alcohol consumption and maintaining a healthy body weight.

The cholesterol-lowering guidelines exist ostensibly to inform doctors how best to reduce the risk of cardiovascular disease in their patients. If the evidence suggests, as it does, that the most effective measures involve the adoption of healthy lifestyle habits, with statin therapy a very distant secondary consideration, then the guidelines should reflect this. Had this been the case, both Jane and her doctor would have been able to prioritize Jane's approach to protecting her health far more effectively.

STATIN SIDE EFFECTS

As the statin debate continues, one of the major issues has become the potential side effects of the drugs. Although statins have been on the market for more than thirty years and are the most frequently prescribed class of drugs in the United States, the medical community still doesn't have good information on the subject. Our

---

* The U.S. Nurses' Health Study was begun in 1976 to study the effects of oral contraceptives on women, and it has since expanded to examine cancer prevention, cardiovascular disease, type 2 diabetes, and other chronic diseases. Initially, the study included approximately 121,700 women ages thirty to fifty-five, and to date it has added 150,000 more, with the latest subjects ranging from age nineteen to fifty-five. Today, 90 percent of the funding for the studies comes from the federal government.

lack of knowledge is a consequence of the commercial influence on the design of clinical research. One author writing in the *Journal of the American Medical Association* explained: "How could the statin RCTs [randomized controlled trials] miss detecting mild statin-related muscle adverse effects such as [muscle pain]? By not asking. A review of 44 statin RCTs reveals that *only 1 directly asked about muscle related adverse effects*" (italics mine).

The single RCT that proactively called study participants (biweekly over a six-month period) to ask about those adverse effects found that 4.8 percent more people treated with a statin than placebo experienced muscle symptoms. Because the participants in this short-term study were young — average age forty-four — and in good health, these results almost certainly underestimate the frequency of statin-related symptoms experienced by the older and sicker people who typically take these drugs for many years. In 2015, the European Atherosclerosis Society issued a consensus statement, relying on data from registries and observational studies; it reported that statin-related muscle symptoms occurred in 7 to 29 percent of people taking statins.

Other side effects that have been associated with statins include liver dysfunction, acute renal failure, cataracts, cognitive symptoms, neuropathy, sexual dysfunction, decreased energy and exertional fatigue, and psychiatric symptoms such as depression, memory loss, confusion, and aggression. But relatively little scientific attention has been focused on this aspect of the statin controversy, so the causal role of statins remains unproven.

HOW TO FIX THE CHOLESTEROL GUIDELINES

Jane's doctor had provided her with impeccable primary care, managing her cholesterol level exactly as recommended by the clinical practice guidelines. But (as will be addressed in part II) neither Jane nor her doctor understood how those guidelines had been developed. They did not know that the drugmakers had designed the studies to produce results that would persuade doctors to prescribe their products; that neither the individual studies nor the CTT meta-analyses had been independently verified by peer reviewers or journal editors; or that the experts who wrote the guidelines, many of whom had financial ties to the industry, were unable to access or independently analyze the underlying data.

Very little progress is being made toward rectifying these violations of the basic requirements of good science. In the case of the statin study data, shortly after their unanimous rejection of Professor Collins's demand for retraction of our article, the *British Medical Journal* editors wrote to the lead investigators of thirty-two major statin trials and asked them to make their study data available for independent analysis. One year later, despite follow-up e-mails and phone calls, only seven of the thirty-two investigators had even bothered to respond. The *British Medical Journal* editors then implored statin researchers to make their data available: "The statins trialists have huge potential influence, and they have a choice. They can take the lead on transparency or be pulled kicking and screaming into the light."

It is just plain scientifically and morally wrong that drug companies are not required to make the data from their clinical trials for statins (and indeed for all drugs) available for independent analysis and review before the results are published as individual studies, incorporated into meta-analyses, or included in guidelines. This is especially true when the ever-growing number of people for whom statin therapy is recommended are at lower and lower risk and therefore derive less and less benefit.

Finally, if the primary goal of the statin clinical trials had been to optimize health rather than simply to sell more statins, drug companies would have randomized people not just into two groups (statin or placebo) but into four groups (placebo, statin, intensive lifestyle modification, or both statin *and* lifestyle modification). Such a study design would have shown whether the best approach to heart-disease prevention was adoption of healthy lifestyle habits, statin therapy, both, or neither. But with statin sales going so well, the drug companies had no incentive to risk asking the more relevant question.

The most important lesson to be learned from Jane's story? Taking a pill to reduce your risk of heart disease is certainly easier than adopting healthy lifestyle habits. And this would be fine if the pills worked well. But they don't. If you really want to decrease your risk of heart attack and stroke and maximize your chances of staying healthy overall, by far your first priority should be maintaining a healthy lifestyle: exercising for a half an hour or more at least five times a week, not smoking, eating a healthy diet, maintaining a healthy body weight, and drinking alcohol in moderation.

Americans and their health-care providers must remember that the real goal is *not* to have the lowest cholesterol level but to have the lowest risk of heart disease and stroke. Had the guidelines Jane's doctor followed so scrupulously with regard to cholesterol levels and statin therapy first called for an *in-depth program* to help Jane overcome her resistance to making the lifestyle changes that would have far more effectively protected her health, surely her doctor would have followed that advice.

Such programs would be the most effective and efficient path toward decreasing the burden of cardiovascular disease in the United States. But, as will be discussed in part II, commercial control of medical "knowledge" focuses our preventive efforts on the use of drugs and downstream interventions rather than on helping individuals and communities maintain healthy lifestyles. That said, many doctors feel pessimistic about their ability to help people make these positive lifestyle changes. But the Diabetes Prevention Program study discussed in the next chapter shows that doctors' pessimism on this matter is clearly not supported by the results of a well-designed study to determine the best way to decrease the risk of developing type 2 diabetes. However, the more important story in the next chapter involves yet another way drugmakers extract money from users of their products, with sometimes fatal results.

4

INSULIN INC.: THE EXPLOITATION OF DIABETES

Nearly a century after its discovery, there is still no inexpensive supply of insulin for people living with diabetes in North America, and Americans are paying a steep price for the continued rejuvenation of this oldest of modern medicines.

— JEREMY GREENE AND KEVIN RIGGS,
New England Journal of Medicine

The first recorded treatment of "excessive urination" was found in an Egyptian papyrus written around 1550 BCE. Fifteen hundred years later, a Greek physician gave the disorder the name *diabetes*, meaning "siphon." Diabetes occurs when the body loses its capacity to make insulin (the hormone that regulates the movement of glucose from the bloodstream into cells) or becomes resistant to the effect of normally produced insulin. The first situation results in type 1 diabetes mellitus; the second, far more common, results in type 2 diabetes mellitus (often referred to as T1DM and T2DM, respectively).

Over the past twenty years, the number of adults in this country diagnosed with diabetes has more than doubled. In 2018, according to the Centers for Disease Control and Prevention, 26.9 million Americans had been diagnosed with diabetes, and 7.4 million of them were being treated with insulin. How insulin treatment is managed is thus an important issue for patients, prescribers, insurers, and drugmakers. This chapter addresses three aspects of insulin therapy: the elegance of the scientific innovation, industry's support of scientifically unsupported standards of diabetes care, and the unconscionably high price of insulin in the United States.

All three of these aspects contributed to the tragic fate of Alec Raeshawn Smith. Just before his twenty-fourth birthday, Alec was diagnosed with type 1 diabetes. At the time, his mother, Nicole Smith-Holt, was an employee of the State of Minnesota, which offered good health insurance that would cover her son until he turned twenty-six. Even with that coverage, however, Alec's out-of-pocket costs for insulin and diabetic supplies ran between $250 and $300 per month.

For two years, Alec followed a diabetes regimen — long-acting Lantus before bed and rapid-acting Humalog before meals (both members of the latest generation of so-called insulin analogs) — that kept him healthy and active. He was promoted to manager at the restaurant where he worked and earned a good salary, approximately $35,000 a year. Ms. Smith-Holt told me the restaurant owners treated her son like family, but their chain was just too small for them to offer employees health insurance. So after Alec turned twenty-six, he was on his own to cover his medical expenses.

The first time Alec went to purchase insulin without health insurance, the pharmacist told him the monthly cost of his insulin and supplies would be $1,300. Had Alec been told at that point that a far less expensive option was available, the trajectory of the story that follows would have been radically different. But he wasn't, and $1,300 was much more than he could afford. So he did the only thing he could think of: He started "rationing" his insulin — that is, using less than his doctor had prescribed — to save money.

His health went downhill quickly. When Alec's girlfriend visited him in his apartment on a Sunday evening just three weeks after he had lost his health-care coverage, he was complaining of abdominal pain, shortness of breath, headache, and fatigue, symptoms he attributed to constipation. Uncharacteristically, Alec called in sick Monday morning. His girlfriend went to check on him at 5 p.m. the following day. When he didn't come to the door, she called him. She could hear his phone ringing inside the apartment, but there was no response. She climbed in through a window and found him on the floor, dead. The medical examiner's report showed that Alec had died sometime on Monday. The cause of death was determined to be ketoacidosis due to undertreatment of his diabetes.

To understand this diagnosis, it's necessary to know how type 1 diabetes works. It is believed to be caused by autoimmune destruction of the cells in the pancreas that produce insulin, the hormone that regulates glucose metabolism. As insulin production falls toward zero, less and less glucose can pass from the bloodstream into the body's cells to be used as fuel. This causes the level of sugar in the blood to rise and forces glucose-deprived cells to metabolize fatty acids for energy. Ketones that are produced as a byproduct of this process build up in the blood and make it dangerously acidic. At the same time, the elevated blood sugar causes dehydration by pulling too much water out of the body through the kidneys. This is called diabetic ketoacidosis, and it can be rapidly fatal — as it was with Alec — unless treated intensively with intravenous insulin and rehydration, along with careful monitoring and correction of electrolyte imbalances.

Although the immediate cause of Alec's death was listed as diabetic ketoacidosis, the real cause was that the insulin manufacturers had convinced prescribers that the latest generation of insulin was necessary and then jacked up the price to unimaginable levels. And as outrageous as the cost of insulin is for people with type 1 diabetes, it is even more outrageous for people with type 2 diabetes because, for them, the supposed benefits of the newest insulins have been even more grossly misrepresented. Understanding how this came about requires an examination of those insulin-therapy issues, beginning with how insulin technology has changed over the past forty years and how the price has changed over the past ten.

NEWER DOESN'T NECESSARILY
MEAN BETTER

If we take *innovation* to mean "creation of new technology," then extraordinary innovations have certainly occurred in the manufacturing of insulin therapy over the past one hundred years. But if we assume that the definition of *innovation* includes providing previously unavailable benefits, then the "previously unavailable benefits" of these recent innovations have been largely limited to the profits they have created for the insulin manufacturers.

Insulin therapy for the treatment of diabetes in humans began on January 11, 1922. Leonard Thompson, a twelve-year-old boy whose diabetes (later categorized as type 1) had been diagnosed two years earlier, lay dying in a Toronto hospital. Leonard's father gave permission for him to be treated with an "extract" prepared from the pancreases of dogs. His blood sugar went down by 25 percent, but this was not enough to significantly improve his clinical condition. Twelve days later he began a two-week series of daily injections of a more concentrated pancreatic extract. He improved immediately; his urinary sugar and ketones (a marker of ketoacidosis) decreased to nearly normal levels after the first day of therapy. Leonard lived another thirteen years before dying of pneumonia, a probable complication of his diabetes.

In January 1923, the three scientists who developed the pancreatic extract* sold their patent to the University of Toronto for one dollar each. (At the time, academic researchers frowned on profiting from their discoveries.) By the end of 1923, two of the current giants of insulin production — Eli Lilly and the Danish company that later became Novo Nordisk — had begun commercial production of insulin derived from pig pancreases.

Improvements in the purity and duration of action of insulin

---

* Orthopedic surgeon Frederick Banting, medical student Charles Best, and biochemist J. B. Collip.

were incremental until the 1970s, when a new era of radical scientific innovation began. Recombinant DNA technology had previously been the stuff of science fiction, but in 1978 scientists from Genentech and City of Hope National Medical Center successfully inserted cloned human DNA fragments that coded for human insulin into the genes of *E. coli* bacteria. This process allowed vats of *E. coli* to become factories that produced insulin with exactly the same chemical structure as human insulin. Genentech licensed the technology to Eli Lilly, and four years later, in 1982, recombinant human insulin became the first genetically engineered drug approved by the FDA. Short-acting recombinant human insulin, Humulin R (*R* for "regular"), was approved for use before meals, and intermediate-acting recombinant human insulin, Humulin N (*N* for "neutral protamine Hagedorn," or NPH, a reference to its pH and its creator, Hans Hagedorn), was approved for blood-sugar control for up to twenty-four hours.

Insulin was the perfect candidate to become the first genetically engineered drug—the benefit of insulin was long established, and the claims that genetically engineered human insulin was purer than insulin from pigs and cows had prima facie appeal. For example, a 1984 advertisement in a medical journal showed a picture of a young boy sitting on a couch, teddy bear on one side and robot warrior on the other, injecting himself with animal insulin. The caption tugged at readers' heartstrings: "He's four years old. And already he's living in the past." Lilly claimed its human insulin was "outstandingly pure. And it's a less immunogenic form of insulin than that which comes from the pancreas of pigs and cattle." What sort of cad would deprive this diabetic child of a type of insulin that was newer, purer, and less likely to evoke an unwanted immune response?

As a practicing doctor, I remember being told that human insulin was less likely than animal insulin to cause local reactions in patients. I was dubious; animal insulin was working just fine, so switching my patients to human insulin would just increase the cost and require extra visits to adjust the dosage. Besides, if any of my patients had developed a reaction to animal insulin, I could have switched them to the newer insulin at that point. But the experts insisted we doctors should switch, so switch we did.

In 1996, the FDA approved the first of the second generation of bioengineered insulins, called insulin analogs, which drug companies claimed were even better than human insulin. The first analog, rapid-acting insulin lispro (brand name Humalog), differed from human insulin by a single amino acid; this created a form of insulin that more closely mimicked the normal nondiabetic insulin response to eating. Humalog was soon followed by insulin aspart (Novo Nordisk's Novolog) and insulin glulisine (Sanofi's Apidra), both with similar claims of improved blood-sugar control.

In 2000, the first of the long-acting analogs entered the market, insulin glargine (brand name Lantus); it was followed five years later by insulin detemir (Novo Nordisk's Levemir). The manufacturers claimed their long-acting analogs provided more even effects over twenty-four hours than intermediate-acting recombinant human insulin did. As a result, they said, the number of insulin-associated hypoglycemic episodes (when blood sugar drops below normal) would decrease.

Manufacturing insulin with recombinant DNA technology was heralded as a major scientific breakthrough. Even so, each successive generation of bioengineered insulin should not have been adopted as the treatment of choice without its clinical superiority having been proven.

The first of these transitions was from insulin derived from pigs and cows to bioengineered insulin that had exactly the same molecular structure as human insulin. The advantages appeared obvious. By 2000, this presumed progress was reflected in the fact that 99 percent of the animal insulin once used in the United States had been replaced by recombinant human insulin. But — to ask what seemed like a dumb question at the time — did this radically innovative insulin really offer any meaningful clinical advantage over the animal insulins? A 2003 review of all relevant studies by the Cochrane Collaboration (a nonprofit organization that performs comprehensive systematic reviews to determine optimal therapy, discussed further in chapter 6) found no "clinically relevant differences" between the two types of insulin for people with type 1 or type 2 diabetes. The review concluded with a stern and prescient warning: "The story of the introduction of human insulin might be repeated by contemporary launching campaigns to introduce pharmaceutical and technological innovations that are not backed up by sufficient proof of their advantages and safety."

Indeed, a big clue to understanding the push behind this scientifically unsupported transition was the low cost (and therefore low profit potential) of a vial of that old-fashioned, supposedly impure and immunogenic animal insulin. Sixty years ago, one vial of animal insulin cost 84 cents, the equivalent of $7.44 in 2021 dollars. Newer might not be clinically superior, but it gave the manufacturers a chance to bring in a whole lot more revenue. To be clear, having another type of insulin on the market as a second-line choice for patients experiencing difficulty with the animal insulins was a good backup option. But that's not at all what was happening.

Three Cochrane Reviews published between 2006 and 2008 compared the benefit and harm of the second generation of bioengineered insulin, the insulin analogs, to the recombinant human insulins. Although the insulin analogs were touted by the pharmaceutical companies as more closely reproducing the body's natural insulin activity, the reviews concluded that for people with type 1 diabetes, the insulin analogs provided only slightly better control of blood sugar (deemed "clinically unremarkable"). The reviews also found only a small reduction in the risk of hypoglycemic events — one fewer hypoglycemic event severe enough to require the help of another person for every five years of treatment with analog rather than human insulin. And for people with T2DM, the reviews showed that the human insulins, compared to the insulin analogs, were "almost identically effective."

The general conclusion was that the evidence warranted "a cautious response to the vigorous promotion of insulin analogues." Nevertheless, like the previous transition from animal to recombinant human insulin, the response was anything but cautious; by

2010, over 90 percent of insulin prescriptions in the United States for people with T2DM were for insulin analogs rather than recombinant human insulin.

One might take the position that, all other things being equal, why not switch to the newer insulins? But all other things were far from equal. The three manufacturers controlling 99 percent of the global insulin market (Sanofi, Novo Nordisk, and Eli Lilly) were able to act like a cartel. This was especially true in the United States, where the passage of Medicare Part D allowed senior citizens to purchase prescription drug coverage starting in 2006, but the program *specifically prohibited the government from negotiating price* with the manufacturers. And in 2010, Obamacare legislation *specifically prohibited the consideration of cost-effectiveness* in coverage decisions for government programs or guideline recommendations. (More on these issues in chapters 8 and 9.)

The cost of insulin analogs is exhibit A in understanding our out-of-control drug prices. When Lilly first launched its short-acting insulin analog Humalog in 1996, the price was $21 per vial. Between 2012 and 2018 — after convincing the nation of the superiority of insulin analogs — insulin manufacturers raised the price of the drug in the United States *more than 18 percent each year.* By May 2017, Lilly had increased the undiscounted retail price to $330 per vial. To put the extreme pricing of insulin analogs in the United States in perspective, had Alec lived in Canada, he could have purchased that vial of Humalog for $38 (with similar prices on other insulin analogs) and would almost certainly be alive now. In 2019, at discounted retail prices, the cost of one year of therapy with the insulin analogs peaked at an average of $5,224, assuming a ratio of 70 percent Lantus ($286/vial) to 30 percent Humalog ($178/vial) and an average use of 20.6 vials per year. In contrast, treatment with recombinant human insulin cost just $468 per year.*

Was spending eleven times more — an extra $4,756 per year — for

---

* The discounted retail price for human insulin was $28.38 per vial, with an average of 16.5 vials required per year.

the (at best) small advantage of the insulin analogs over recombinant human insulin cost-effective for people with type 1 diabetes? The answer for Alec Raeshawn Smith was, in retrospect, clearly no, and he was by no means alone. The same year that Alec died, one out of every four people receiving care at the Yale Diabetes Center in New Haven, Connecticut, "reported cost-related insulin under-use," and under-users were three times more likely to have poor control of their blood sugar.

Why do American doctors keep prescribing insulin analogs for people with type 1 diabetes who cannot afford them? The answer lies in the expert recommendations doctors received from guidelines and position papers issued by trusted organizations like the American Association of Clinical Endocrinologists (AACE) and the American Diabetes Association (ADA). (More on this later.) These authoritative recommendations — calling for "the use of insulin analogs for most patients with [type 1 diabetes]" — combined with the relentless price hikes represented not a clinical advance but a death sentence for Alec Raeshawn Smith.

People with type 1 diabetes need insulin to survive, but only about one out of every twenty Americans with diabetes has type 1. Almost all the rest have type 2 diabetes mellitus, which is caused not by an abrupt *reduction in the amount* of insulin being produced, as in type 1, but by the cells of the body gradually becoming *resistant to the effect* of insulin. Early in the course of T2DM, the pancreas can compensate by increasing insulin output, but as insulin resistance increases, the pancreas cannot produce enough insulin to keep up, and blood-sugar levels rise. Symptoms develop gradually; increased glucose in the blood pulls too much water out of the body, causing frequent urination, increased thirst, increased hunger, fatigue, blurred vision, and weight loss. Extremely elevated blood sugar in people with type 2 diabetes can cause serious problems but rarely results in the deadly ketoacidosis that can occur in people with type 1.

The risk factors for developing type 1 diabetes are not known (beyond a small increase in risk associated with a parent or sibling having type 1 diabetes), but we do know the primary risk factors for T2DM: obesity, which raises the risk at least sixfold, and poverty, which increases the risk two- to threefold. Adopting healthy lifestyle habits is the most effective way to prevent and treat T2DM (more on this later), but if this fails, the first blood-sugar-lowering medication prescribed is usually the inexpensive oral medication metformin. If this does not adequately control blood sugar, another oral or a non-insulin injectable medicine may be added. All people with type 1 diabetes require insulin at the time of diagnosis whereas only about one-quarter of people with T2DM eventually require insulin. But because so many more Americans have been diagnosed with type 2 diabetes (25.4 million) than with type 1 (1.5 million), approximately four-fifths of all the insulin prescribed in the United States is used to treat T2DM.

Clearly, there was a lot of money to be made by ensuring that insulin analogs rather than recombinant human insulin became the standard for people with T2DM who required insulin therapy. And the United States — with its lack of regulation of drug prices, preference for the newest drugs, and highest rates of obesity and diabetes among wealthy countries — has been by far the best place for drug companies to do that. For people with T2DM, the insulin analogs have not been shown to provide an advantage over the human insulins, so any higher cost — whether double the price or ten times the price — is unjustified. Countries that allowed cost-effectiveness to be integrated into their coverage recommendations understood this. The UK's National Institute of Clinical Excellence recommended human insulin (not insulin analogs) as first-line therapy for people with type 2 diabetes who required insulin. So did New Zealand's Pharmaceutical Management Agency. And the Canadian Agency for Drugs and Technologies in Health. And the German Institute for Quality and Efficiency in Health Care. Given the lack of scientific evidence supporting the superiority of insulin analogs over human insulin in treating T2DM, the manufacturers faced the difficult challenge of finding a way to push these drugs in the United States. Ultimately, this was accomplished. Using the channels doctors trusted, marketers persuaded them to prescribe this far more expensive — but no more effective — class of drugs.

INDUSTRY INFLUENCE ON T2DM STANDARDS OF CARE AND PRACTICE GUIDELINES

In 1997 the Diabetes Quality Improvement Project (DQIP), founded by the Centers for Medicare and Medicaid Services, the American Diabetes Association, and the National Committee for Quality Assurance, established the first national standards for diabetes care. The DQIP determined that hemoglobin A1c (HbA1c) levels of 9.5 percent or greater* (the point at which many people experience diabetic symptoms) would be defined as "poor control." But lacking both the scientific evidence to determine optimal control and the data to determine the needs of the specific populations being monitored, they refrained from establishing a standard level for "good control."

This left the insulin makers with the seemingly insurmountable challenge of convincing doctors that, despite the lack of evidence, prescribing the more expensive insulin analogs was in the best interests of their T2DM patients. That the manufacturers succeeded in persuading doctors to prescribe what now amounts to more than *$20 billion per year* of insulin analogs for people with T2DM shows that effective marketing can have far greater influence than

---

* The diagnosis of diabetes is made when a patient's HbA1c level is 6.5 percent or higher, a fasting blood sugar is above 125 mg/dL, or a random (non-fasting) blood sugar is 200 mg/dL or above. HbA1c measures the percentage of hemoglobin molecules in red blood cells that have bonded to glucose. The higher the blood sugar, the greater the percentage of hemoglobin molecules bound to glucose. Because the life span of a red blood cell is around three months, the HbA1c level provides an estimate of the average blood-sugar level over the past three months.

scientific evidence on the way physicians make prescribing decisions.

The insulin makers used a two-step strategy. First, they needed to establish standards of diabetes care that called for tight control of blood-sugar levels (meaning close to normal nondiabetic levels) to decrease the risk of serious complications, notwithstanding the lack of scientific evidence to support this claim; this would allow the manufacturers to highlight the more precisely timed effect of insulin analogs. And second, drugmakers needed clinical practice guidelines recommending insulin analogs as a way to achieve this tighter control with less risk of hypoglycemia, notwithstanding the lack of evidence supporting that claim.

In 2002 the insulin manufacturer Aventis (soon to be acquired by Sanofi) took the setting of the standards of diabetes care into its own hands by funding the Aim-Believe-Achieve campaign. The public relations and marketing firm Burson-Marsteller was hired to develop the program. Instead of recommending that HbA1c levels be kept to below 9.5 percent as recommended by the DQIP, Aim-Believe-Achieve called for HbA1c levels below 7 percent. Evidence that this lower threshold would reduce the risk of cardiovascular disease or other major complications for people with T2DM did not exist. The only ones certain to benefit from doctors believing that expensive insulin analogs would help their T2DM patients achieve this goal more effectively and safely than recombinant human insulins were the makers of those insulin analogs.

The following year, the Bridges to Excellence (BTE) initiative was established. Its goal was to encourage doctors to "improve" the quality of their medical care, and this would be accomplished by rewarding physicians who met the so-called quality standards defined by BTE. One of those standards was maintaining HbA1c levels at or below 7 percent in at least half of the diabetic patients being treated. Those physicians who were among the "top performers" in meeting this standard qualified for a reward (called "pay-for-performance") of up to $80 per patient per year. The BTE diabetes module was cosponsored by the National Committee for Quality Assurance (NCQA), the largest accreditor of outpatient health plans, and the American Diabetes Association. Both of these non-profit organizations are supported by industry funding.

The unsubstantiated treatment standard of an HbA1c of 7 percent or less was given a big boost by prominent review articles in the two most influential American medical journals. A 2003 Clinician's Corner article in *JAMA* reported that insulin analogs for the treatment of both type 1 and type 2 diabetes allowed tighter control of blood sugar "associated with fewer episodes of hypoglycemia." Similarly a 2005 Drug Therapy review article published in the *New England Journal of Medicine* posited the benefit of a target HbA1c of 7 percent or less, based on the claim that "near-normal [HbA1c] levels reduce the risk of diabetic complications."* This article similarly made the link between tight control of blood sugar and the benefit of insulin analogs: "Insulin analogues have met this demand ... with an attendant decrease in the risk of hypoglycemia." One author common to both articles reported having received consulting fees and honoraria from Aventis, Novo Nordisk, and Eli Lilly.

In 2006, the NCQA, still without evidence of the clinical benefit of tight control for people with T2DM, adopted the even more stringent standard that not just half but *all* patients with diabetes should have HbA1c levels of less than 7 percent. Unbeknownst to almost all members of the NCQA's Committee on Performance Measurement who voted to adopt this new standard, the technical-expert panel of a parallel quality-assurance organization — the National Diabetes

---

* The author of this article relied on the results of the UK Prospective Diabetes Study (UKPDS) to support this claim. But the UKPDS had actually shown that intensive blood-glucose control did *not* reduce the risk of the major complications of diabetes: death, heart attack, stroke, kidney disease, and amputations. The only advantage of tight control was a slightly lower need for retinal photocoagulation, a laser treatment to prevent blindness, sometimes a complication of diabetes. But even though less photocoagulation was required in the UKPDS study, the people who received the intensive blood sugar-lowering therapy were no less likely to develop blindness than those receiving the conventional, nonintensive therapy.

Quality Improvement Alliance — that had access to the same scientific evidence *unanimously rejected* the quality standard of HbA1c of less than 7 percent as far too low. The alliance had proposed lowering the earlier "poor control" standard of not above 9.5 percent to not above 9 percent, a far less stringent standard than the NCQA's.

Writing in the *Journal of the American Medical Association,* David Aron and Leonard Pogach (physicians with extensive experience with diabetes standards of care and quality improvement) suggested "an industry-driven campaign" had played a significant role in the NCQA's adoption of the new standards for HbA1c levels: "It appears that a bandwagon effect openly funded by industry, may have contributed to the adoption of a measure *despite significant and documented scientific disagreement of opinion*" (italics mine).

Then, in 2008, results from the high-profile, large, and rigorously designed ACCORD (Action to Control Cardiovascular Risk in Diabetes) study were published in the *New England Journal of Medicine.* The findings threw a big wrench into the movement for tight control of blood sugar. Or at least they should have. The study had been designed to determine whether tight blood-sugar control reduced heart attacks or strokes for people with T2DM better than standard therapy did. It was stopped prematurely, 3.5 years into the planned 5.6-year study, because the risk of death was significantly higher (22 percent) in people treated with intensive therapy to achieve tight control,* and three times as many people in the tight-control group had experienced a severe hypoglycemic event (10.5 percent versus 3.5 percent). The trial provided definitive evidence that intensive treatment for T2DM was harmful. NCQA quickly put its goal of a hemoglobin A1c of less than 7 percent for all T2DM adults on hold. In 2009, its updated quality measure for glucose control was far less stringent, calling for "at least 60% of patients to have an HbA1c value of [less than] 8%."

As of 2018, five large long-term randomized trials had evaluated

---

* HbA1c levels averaged 6.4 percent in the intensive-treatment group and 7.5 percent in the standard-therapy group.

the benefits of tight glucose control versus less intense control* in people with T2DM. Collectively the studies showed that intensive control of blood sugar to a hemoglobin A1c below 7 percent does *not* reduce death, heart attack, stroke, loss of vision, renal disease requiring dialysis, or painful diabetic neuropathy. At the same time, all five studies showed that more intensive treatment led to more complications, like hypoglycemia and even death in the AC-CORD study. Notwithstanding all this evidence, in 2019 all but one of the U.S. professional organizations making recommendations for blood-sugar control still called for an HbA1c below 7 percent for most adults with T2DM. The single exception was the American College of Physicians, which called for an HbA1c level between 7 and 8 percent.

Until recently, two of the most influential organizations in diabetes care recommended insulin analogs over recombinant human insulin for people with T2DM. Through 2017, the consensus statement from the AACE on the treatment of T2DM called for the use of insulin analogs — both long- and short-acting — over recombinant human insulin. In 2018 that consensus statement was updated; it still recommended both long- and short-acting insulin analogs over recombinant human insulin but conceded that "for those who find the more costly analog insulins unaffordable, human regular insulin or premixed human insulin for T2D are less expensive options." The acknowledgment that people who can't afford or don't want to waste almost $5,000 a year will do just fine with recombinant human insulin is important, given that 39 percent of Americans do not have the funds to cover an unexpected expense of $400.

Similarly, through 2018, the American Diabetes Association, in its Standards of Medical Care in Diabetes, recommended insulin analogs over recombinant human insulin for people with T2DM. In 2019, the ADA changed its recommendations. For long-acting insulin, the ADA wrote, "Control of fasting glucose can be achieved

---

* HbA1c levels between 6.3 and 7.4 percent versus HbA1c levels between 7.3 and 8.4 percent.

with human NPH insulin or with the use of long-acting insulin analog," and it conceded that the purported lower risk of hypoglycemia associated with long-acting insulin analogs compared to human NPH insulin was "generally modest and may not persist." For rapid-acting insulin, the ADA wrote, "Meta-analyses of trials comparing rapid-acting insulin analogs with human regular insulin in patients with type 2 diabetes have not reported important differences in [HbA1c] or hypoglycemia." In other words, the ADA finally acknowledged, after many years of recommendations and many tens of billions of dollars wasted on insulin analogs for people with T2DM, that the earlier generation of recombinant human insulin worked just as well.

The guidelines of both organizations declared financial ties to many drugmakers. The AACE acknowledged in its consensus statement the "generous support" of twenty-seven corporate donors that made diabetes-related drugs, tests, or other products. In addition, nineteen of the twenty-one authors of its consensus statement, including the chair, declared financial ties to companies that produced drugs or related products for diabetes. Likewise, in 2019, nine of the fourteen members of the committee responsible for producing the American Diabetes Association's Standards of Medical Care in Diabetes had financial ties to diabetes drug or device manufacturers. I am not accusing the conflicted experts on the AACE and ADA panels of being bought by these drug companies, but it is a safe bet that the corporations would not have been giving money to experts who recommended *decreasing* the use of their most profitable products.

In retrospect, it is obvious what caused doctors to accept the scientifically unsubstantiated belief that treating their T2DM patients with far more expensive insulin analogs was in their patients' best interest. The problem is that, in real time, practicing physicians cannot get accurate and balanced scientific evidence about new drugs; meanwhile, commercial interests have the resources to overwhelm health-care providers with self-serving and unsupported claims of the efficacy of their expensive drugs.

For patients with type 1 diabetes, like Alec, the long-acting insulin analogs "probably" provide an advantage over recombinant human insulin. But the difference is not black and white, and the shades of gray grew progressively darker as the price of insulin analogs rose in the years before Alec's death. Had Alec's doctor been accurately informed about the limited additional benefit of insulin analogs — instead of being bombarded with commercially influenced medical journal articles, guidelines, and standards of care — he would have been able to help Alec assess the pros and cons of switching to recombinant human insulin, even temporarily, when he lost his insurance coverage.

In 2020 (sadly, too late for Alec), the American Diabetes Association finally acknowledged that for some people with type 1 diabetes, "despite the advantages of insulin analogs," their cost might be prohibitive. For these patients, the ADA stated, "there are multiple approaches to insulin treatment" and "some form of insulin" — including human recombinant insulin — should be used to keep patients healthy. The ADA was belatedly acknowledging that the differences between human recombinant insulin and insulin analogs were far less important than keeping patients with type 1 diabetes out of ketoacidosis. The moral of this story is that doctors had been sold a bill of goods about the advantages of insulin analogs when what they needed was accurate and complete scientific information. How to rectify this situation is the subject of part III of this book.

A RATIONAL APPROACH

While patients with T2DM are wasting approximately $23 billion per year* on expensive but not clinically superior insulin ana-

---

* Approximately 5.9 million Americans with T2DM use insulin, and the difference in price between insulin analogs and recombinant human insulin was almost $5,000 per year in 2019. The $23 billion excess expenditure assumes

logs (and insulin makers and their investors are getting richer and richer), the United States is in the midst of a rapidly worsening epidemic. The number of Americans diagnosed with diabetes has increased fourfold since 1982, from 5.7 million to 26.9 million, with almost all of that increase due to T2DM. The United States now has the highest percentage of adults with diabetes among the wealthy nations. The cause of our epidemic of T2DM is not obscure — the increasing rate of diabetes almost exactly parallels the increasing rate of obesity. Between 1960 and 1980, the percentage of obese American adults went up only slightly, from 13.4 to 15 percent. But from 1980 to 2016, the rate of obesity increased from 15 to 39.8 percent. Today, no other country in the Organisation for Economic Co-Operation and Development (OECD)* has a higher obesity rate; ours is fully twice the average of the other countries'. And it continues to grow: by 2030 almost half of American adults (48.9 percent) will be obese.

The definitive study showing how best to reduce the incidence of T2DM was done in the 1990s. The publicly rather than commercially funded Diabetes Prevention Program Research Group randomly assigned more than three thousand people at high risk of developing diabetes to one of three groups: standard lifestyle recommendations plus placebo; standard lifestyle recommendations plus metformin (a blood-sugar-lowering medication); or intensive counseling about healthy diet, exercise, and behavior modification plus placebo. The primary outcome was the difference in the percentage of people who developed diabetes in each group after three years.

The study was highly unusual in that its goal was to determine the best way to decrease the risk of diabetes, instead of just trying to prove the efficacy of a particular drug. Its unusual design tested the

---

90 percent of people with T2DM are currently using insulin analogs, and 90 percent of them could be successfully switched to recombinant human insulin.

* An association of wealthy countries dedicated to promoting economic progress and world trade.

widely held belief that people can't be helped to adopt and sustain healthy lifestyle habits. Indeed, the participants, all of whom had prediabetes* and two-thirds of whom were obese, would have been expected to have significant difficulty with making lasting lifestyle changes.

The results were stunning. First, they resoundingly debunked the conventional wisdom that people won't change their habits and that physician recommendations and programs that attempt to get them to do so are a waste of time. At the end of three years, the people assigned to the intensive-lifestyle-modification group had lost an average of twelve pounds and significantly increased their physical exercise compared to the two other groups. And second, the study showed that by far the best way to prevent diabetes is through positive lifestyle changes. Intensive counseling on lifestyle modification reduced the risk of developing diabetes by 58 percent compared to the placebo group and by 39 percent compared to the group treated with metformin. It is important to note that T2DM is not just a problem of high blood sugar; the underlying disease mechanism causes a doubling of the risk of heart attack and stroke and is the leading cause of kidney failure, adult blindness, and non-trauma-related foot and lower-leg amputations.

Looked at another way, the lifestyle-modification program delayed diabetes by an average of eleven years in this high-risk population, and the cost for each year of healthy life† gained was only

---

* *Prediabetes* is currently defined by a fasting blood sugar between 100 and 125 mg/dL without blood-sugar-lowering medication. The threshold for participation in the Diabetes Prevention Program study was slightly different: a fasting blood sugar between 95 and 125 mg/dL.

† A quality-adjusted life-year (QALY) is a measure of survival that incorporates both the length and quality of life, typically related to a medical intervention. One QALY could be one added year of life in perfect health or two added years of life adjusted for health status allowing 50 percent of optimal quality of life. In the current instance, the cost input is the amount of money required to provide intensive lifestyle counseling to enough people so that the total gain for the group is one QALY.

$1,100. Compared to what other therapeutic interventions typically cost in America to provide a year of healthy life — up to $150,000 or more for each year gained — the cost of the health benefit of intensive lifestyle modification is an enormous bargain.

The Nurses' Health Study (described in the previous chapter) showed an even more dramatic reduction in the risk of developing diabetes associated with adherence to just three healthy lifestyle habits. Female nurses who had a body mass index of less than 25 (that is, who were not overweight), exercised for at least a half hour daily, and ate a healthy diet (one low in trans fats and sugary foods and high in fiber and ratio of polyunsaturated to saturated fat) had an 88 percent lower risk of developing diabetes than nurses who did not maintain these three healthy habits.*

Between 1996 and 2013, the cost of caring for Americans with diabetes rose more than for any other disease category. Two-thirds of that increase, approximately $44 billion annually, was due to rising pharmaceutical expenditures, and a large part of that was due to increased spending on insulin. An optimal approach to minimizing both the health and economic burden of T2DM in the United States would invest our private and public research funds proportionally to produce the greatest health benefit. This would include building on the Diabetes Prevention Program study, which showed that lifestyle-counseling programs could cut in half the number of Americans developing diabetes each year, from two million to one million (and greatly improve general health as well). The cost of implementing community-based healthy-lifestyle programs (like those offered by the YMCA) would be $300 per person per year. The total cost for enrolling all eighty-eight million Americans with prediabetes in such programs would be $26.4 billion per year. Of

---

* Unlike the Diabetes Prevention Study, which was a randomized trial, the Nurses' Health Study is observational and leaves open the possibility that other confounding factors played a role in the observed relationship between maintaining a healthy lifestyle and the risk of diabetes. Nonetheless, paired with a completed randomized trial that strongly supports the findings, it's helpful as a way of quantifying the role of individual habits.

course, this is an enormous amount of money to add to our national medical expenditures. But it is not much more than the $23 billion that would be saved each year by switching 90 percent of the people with T2DM currently using insulin analogs to human insulin.

Now some good news: The CDC is slowly initiating just such a program. Between 2012 and 2016, thirty-five thousand people who had prediabetes or who were otherwise at high risk of developing T2DM were enrolled in the National Diabetes Prevention Program. One-year results on the first fifteen thousand enrollees showed that participants were exercising an average of 150 minutes per week and had lost an average of 4.2 percent of their body weight. The study concluded that the program achieved "promising early results" and that more study on maintaining the duration of session attendance would be helpful.

Supporting community-based diabetes-prevention programs and investing in public health approaches to diabetes prevention are crucial to slowing our nation's diabetes epidemic. But probably the most important intervention is providing doctors with the scientific evidence they need to help their patients make positive lifestyle changes and help communities facilitate those changes. Especially for T2DM, that approach would be much more effective and much less expensive than what we are doing now.

And for the many patients with type 1 diabetes unable to afford the rising cost of their insulin, the most obvious remedy would be to bring the price of insulin in the United States down to the level charged in other wealthy countries. But given the drug companies' power and resiliency, this is unlikely to happen. So plan B would be to make sure doctors understand that continuing to prescribe insulin analogs instead of far less expensive recombinant human insulin is doing those patients much more harm than good. Had the ADA instructed physicians to inquire about their patients' ability to afford insulin analogs prior to 2020, Alec's doctor could have helped him switch to less expensive insulin, which almost certainly would have saved his life.

Finally, our tolerance of prolonged patent-protected monopoly or near monopoly of new technologies (like the insulin analogs) prevents the market from establishing fair prices. If a true market existed for generic versions of insulin analogs, the estimated overall price per person — including manufacturing costs, distribution, and profits — might be no more than $133 per year. Currently, the price per person is more than $5,000 per year.

The changes in insulin therapy over the past forty years reflect the changes in American health care, which is now designed to spend enormous amounts of money and effort on innovative treatments for the downstream manifestations of disease while virtually ignoring far more effective and less expensive upstream prevention. Part II explores how our country has changed during those years and how our health and health care have evolved to reflect those changes.

Pharma Means Business

5

AS AMERICAN SOCIETY GOES, SO GOES AMERICAN HEALTH CARE

American health care does a poor job of delivering health but is exquisitely designed as an inequality machine, commanding an ever-larger share of G.D.P. and funneling resources to the top of the income distribution.

— ANGUS DEATON, 2015 Nobel laureate in economics

Part I showed how physicians' knowledge — or, more precisely, their beliefs — about the benefits of specific drugs was hijacked by the drug companies. Part II explores the central paradox of U.S. health care: Despite spending so much more on health care, Americans are less healthy than the citizens of all other wealthy nations, and we continue to fall farther behind.

This chapter describes the societal changes that refocused the mission of publicly held for-profit corporations in the United States. Instead of serving the interests of a broad range of stakeholders, they pivoted to serving the narrow financial interests of investors and executives. The following chapters will show how Big Pharma took advantage of this trend by gaining control of most of the scientific evidence that doctors rely on to determine optimal care of their patients. I'll use the principles of behavioral economics to explain why pharmaceutical marketing works so well, even on an audience of highly trained health-care professionals. And the final chapter of part II will show how American health care's unique reliance on the market is extremely advantageous to powerful corporations, institutions, and investors but turns physicians into their unwitting agents and leaves the public with a critically dysfunctional health-care system.

WHY NOT SPEND MORE ON HEALTH CARE?

A few years ago, I began a lecture at the University of Michigan by noting that the United States spends an *excess* $1.5 trillion annually on health care compared to other wealthy countries. At the time, this was about twice the U.S. federal budget deficit, two and a half times the annual budget of the Department of Defense, or enough to pay off all the outstanding student loans in America in one year. My intent was to shock the audience, many of them future health-care professionals. But an obviously *un*shocked business major promptly raised his hand and asked if this was necessarily a bad thing. "What's more important than our health?" he mused. "If we must pay a higher price, so be it."

Good question. After all, Americans today are living five years longer than we were in 1980. In a now-classic 2006 article published in the *New England Journal of Medicine,* "The Value of Medical Spending in the United States, 1960–2000," Harvard professor of economics David Cutler — adviser to President Clinton and, later, President Obama — and colleagues wrote that new therapies, including drugs, were largely responsible for rising medical costs. Nonetheless, the authors concluded, "the money spent has provided good value" and "the return on spending has been high." And in 2010, Jonathan Gruber, professor of economics at MIT and a key architect of Obamacare, posited that, for those with good insurance, American health care was the best in the world. Moreover, he asserted, only 10 percent of spending was lost to waste, fraud, or abuse.

A 2017 *New York Times* article concluded that the primary driver of increasing health-care costs was not the aging of the population, as was often thought, but rather the use of new medical technologies. Relying on that 2006 *NEJM* article, the *Times* reported that "the vast majority of the seven years of life expectancy gains in the latter half of the 20th century was due to better — and more costly — treatments for premature infants and cardiovascular disease."

Perhaps the business student had it right. We're living longer. We have the best and newest technology. We are paying a lot for it, yes, but to paraphrase an old hair-dye commercial, we're worth it!

But this attempt to justify our uniquely high spending does not hold up to closer scrutiny. To see why, let's first examine that excess $1.5 trillion we spend annually on health care. Back in 1980, the United States spent 8.2 percent of GDP* on health care, virtually the same as the second-highest-spending country (Germany, at 8.1 percent) and just 1.7 percent above the average of ten comparable countries.† But, as discussed in the introduction, by 2019 American health-care spending had skyrocketed to 17.7 percent of U.S. GDP. At 7 percent more than the average of ten other wealthy countries, this adds up to $1.5 trillion in *excess* spending.

WHAT OUR EXCESS HEALTH-CARE
SPENDING BUYS

In 1980 — when U.S. health-care spending was just slightly above the average of ten comparable countries — the age-adjusted death rate‡ for Americans was lower (better) than the average of those

---

* Comparing countries' level of health-care spending is more fairly based on percentage of GDP because the cost of delivering care, from labor to supplies to transportation, is greater in higher-GDP countries. The impact of higher GDP inflates the comparative per person health-care costs in countries with higher GDPs.

† Australia, Austria, Belgium, Canada, France, Germany, Japan, the Netherlands, Switzerland, and the UK.

‡ To eliminate the potential bias in comparative death rates caused by age differences between countries, death rates can be compared by calculating

other countries. That year, 88,000 *fewer* Americans died than predicted by the average death rate in the other countries. But by 2017, despite our health-care spending having surged to extreme outlier status, our death rate had *increased* so far beyond the others' that *488,000 more* Americans were dying each year — more than 1,300 deaths every day in excess of the number predicted by the average death rate in the other countries.* One would think that more than a thousand excess American deaths every day would, like the toll of coronavirus victims, be headline news every day, but there is very little public discussion of the inferiority of Americans' health.

Worse yet, for the first time in decades, and alone among comparable nations, the lifespan of Americans actually *decreased* from 2014 to 2018. This decline in longevity is particularly noteworthy because it occurred during the second half of the longest period of economic expansion in modern times *and* after the implementation of Obamacare, which, starting in 2014, extended health-care coverage to almost twenty million previously uninsured Americans.

But perhaps longevity is just too crude a measure to capture the benefit of our higher health-care spending. "Healthy life expectancy," or years lived in good health, combines longevity and health-related quality of life.† On that scale, not only had we dropped to sixty-eighth in the world by 2019, but shockingly, the

---

age-specific mortality rates in proportion to the World Health Organization standard population.

* In 2017 the age-adjusted mortality rate for the United States was 840.2/100,000 compared to 690.6/100,000 for the comparable country average. To calculate the number of excess deaths in the United States: [(840.2–690.6)/100,000] x 326,700,000 = 488,743 excess American deaths per year.

† Healthy life expectancy may reflect improved quality of life due to medical care (like improved vision after a cataract operation) that is not captured in the simple measure of longevity. To determine healthy life expectancy, total life expectancy is adjusted for years spent with illness that detracts from quality of life. For example, if someone lived to be eighty-six but had a chronic disease that compromised his quality of life by 50 percent for the last six years of his life, healthy life expectancy would be 86 – (6 x 0.50) = 83.

only countries in the world in which pre-pandemic healthy life expectancy had *decreased* more than in the United States since 2010 were Syria, Yemen, and Venezuela.

Often the comparatively poor health status of people in the United States is assumed to be a reflection of racial and economic disparities, which compromise the health of the less fortunate among us but do not apply to those with more education, good health insurance, and adequate financial resources. This is only partly true. Americans of color and those living in or close to poverty definitely have poorer health, and that is an injustice that demands remediation. But the privileged among us have not been spared. According to the *Journal of the American Medical Association*: "Even non-Hispanic white adults or those with health insurance, a college education, high incomes, or healthy behaviors appear to be in worse health (e.g., higher infant mortality, higher rates of chronic diseases, lower life expectancy) in the United States than in other high-income countries."

This was confirmed by another study that showed that U.S. residents in the highest third of income and education are more likely to have diabetes, hypertension, and heart disease than their English counterparts, despite the fact that the UK spends only 40 percent as much as we do on health care.

The graph shown on page 88, from the Commonwealth Fund, provides a snapshot of the poor performance and higher spending of U.S. health care compared to ten other wealthy countries.* Note that the United States is alone in the lower right quadrant; its spending is far greater and its health-system performance is much worse than those of other countries. This disconnect raises the question: Is the new technology that is driving up the cost of our health care providing enough value to justify its cost?

---

* Performance is calculated for each country as the mean of five domains: care process, access, administrative efficiency, equity, and health outcomes.

Health Care System Performance Compared to Spending

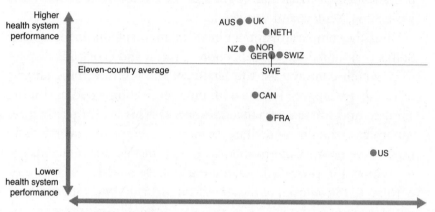

Health-system performance compared to spending in eleven wealthy nations *Eric Schneider et al., "Mirror, Mirror 2017: International Comparison Reflects Flaws and Opportunities for Better U.S. Health Care." Courtesy of the Commonwealth Fund.*

HOW MUCH DOES NEW TECHNOLOGY REALLY HELP?

The drugs discussed in part I — Vioxx, Neurontin for off-label use, statins for people at low risk of cardiovascular disease, and insulin analogs to treat people with type 2 diabetes — are examples of money spent on new drugs that provided little or no benefit. There are many, many other examples of health-care spending that do not provide the value claimed. For instance, the 2006 *NEJM* article mentioned earlier reported that almost 90 percent of the improvement in Americans' longevity between 1960 and 2000 was due to reductions in the rate of deaths from cardiovascular disease and infant mortality. The authors assumed that 50 percent of these improvements in longevity were due to better medical care and were in large part the "result of the development and widespread use" of expensive technological innovation. If this was true, concluded the *NEJM* article, it would provide strong evidence that "the return on [medical] spending has been high." I agree, so let's take a closer look.

The good news is that U.S. infant mortality dropped by almost 80 percent from 1960 to 2015.* But the bad news is that since 1990, the death rate for infants improved almost twice as fast in the other OECD countries. By 2015 the United States ranked a lowly thirty-third from the best in infant mortality among OECD countries.

A comparison of pregnancy care in the United States and three other wealthy countries in the OECD (Australia, Canada, and the UK) shows we *do* use more high-tech resources, but—unlike those other countries—we fail to provide no-cost family planning services or pre- and postnatal medical care. Despite our having almost twice as many neonatologists and 69 percent more neonatal intensive-care capacity in proportion to the number of births, the neonatal mortality rate (defined as death within twenty-eight days of birth) was still 34 percent *higher* in the United States than in the other three countries. This is due in large part to our higher rate of low-birth-weight babies (two-thirds of infant deaths in the United States occur in babies weighing less than five and a half pounds at birth). Also, the rate of teen pregnancies is twice as high in the United States as in the other three countries, and the mortality rate is 50 percent higher for infants born to teenage mothers.

The key finding of the study was that, despite our use of so much more technology, the mortality rate for low-birth-weight babies in the United States was not better (lower) than that of the other three countries. In other words, the increased investment in neonatal technology did not reduce the higher mortality rate even among these high-risk, low-birth-weight babies.

So spending more on technology to care for newborns is not the answer. Quite the contrary—our abundance of high-tech neonatal ICUs and specialists consumes resources that could be used to

---

* U.S. infant mortality declined from 25.9 per 1,000 live births in 1960 to 5.9 in 2015.

provide much-needed low-tech public health measures like family planning, preconception planning, and pre- and postnatal services. The authors concluded by calling into question "the effectiveness of the current distribution of U.S. reproductive care resources and its emphasis on neonatal intensive care." More money to treat high-risk newborns certainly makes intuitive sense, but spending money to reduce the number of high-risk newborns in the first place would provide better value.

U.S. infant mortality statistics are, unfortunately, consistent with its death rates for older children (ages one to nineteen). In the 1960s, the death rate for children in the United States was lower than the average for nineteen other developed countries,* resulting in an average of 3,250 *fewer* deaths of U.S. children each year over that decade. But because the death rate for children has been going down so much faster in the other countries than it has in the United States, an *annual average of 16,380 more deaths of American children — totaling 655,200 excess deaths* — has occurred over the past forty years, compared to the average of those other countries. Numbers like these evoke images of war and famine, not the failures of high-cost U.S. health care. Furthermore, the death rate for children in the United States has lost ground in comparison to other countries in each of the past four decades.† The authors concluded that our unique combination of a high poverty rate and a weak social safety net was the key factor making "the US the most dangerous of wealthy nations

---

* Australia, Austria, Belgium, Canada, Denmark, Finland, France, Germany, Iceland, Ireland, Italy, Japan, the Netherlands, New Zealand, Norway, Spain, Sweden, Switzerland, and the UK.

† Between 2000 and 2010, among fifteen- to nineteen-year-olds, deaths related to firearms were 82 times more frequent and deaths related to motor vehicle accidents were 2.1 times more frequent in the United States than in the other nineteen countries, accounting for about two-thirds of the excess American deaths in this age group compared to twenty-six other high-income OECD countries. Even so, the difference in the number of childhood deaths is dominated by our far-greater-than-average infant mortality rate.

for a child to be born into." Spending more on medical technology is not likely to be the most effective or efficient way to solve this problem.

The 2006 *NEJM* article also reported that 70 percent of the U.S. improvement in life expectancy was due to reductions in cardiovascular mortality. Again, the article attributed half of the improvement to new technology. But as with infant mortality, improvement in American cardiovascular health is lagging behind that of comparable countries that spend a lot less on health care. As of 2017, the United States ranked twenty-sixth among OECD countries for mortality from heart attacks and related heart disease, and fifteen other OECD countries had improved as much as or more than we had since 2000. In an interview, Professor Cutler himself provided an important illustration of how this works. He pointed out that the province of Ontario, Canada, and the state of Pennsylvania have about the same population, but Pennsylvania has five times as many hospitals that do open-heart surgery. And yet, despite five times the capacity for open-heart surgery, mortality rates after heart attacks are no better in Pennsylvania than in Ontario. And many more post–heart attack citizens of Pennsylvania have to go through post-surgical pain, rehabilitation, and potential complications of open-heart surgery.

The conclusion that much of the extra health-care spending in the United States is due to expensive new medical technology is certainly established. But why do doctors choose expensive new drugs and other technologies that offer little or no therapeutic advantage over less expensive alternatives? Why do they welcome more neonatal ICUs rather than advocate for more effective pre- and postnatal public health interventions? Why do they support greater capacity to perform open-heart surgeries instead of at least as effective and far less expensive cardiac care that doesn't have the risks associated with major surgery? To answer these questions, we need to look at how our country has changed over the past seventy-five years, particularly with respect to the role of corporations in Americans' health and health care.

AFTER WORLD WAR II: THE (ALMOST) GOLDEN ERA OF AMERICAN CAPITALISM

The American economy grew rapidly in the first three and a half decades after World War II, and that growth was shared with working- and middle-class Americans. FDR's New Deal brought Social Security, unemployment insurance, unions and collective bargaining, minimum-wage laws, and the Home Owners' Loan Corporation (to prevent Depression-era home foreclosures). Industrialization for the war effort and public investment in infrastructure led to full employment and economic growth. Corporations addressed the needs of a broad range of stakeholders — consumers, employees, the community, and the nation in addition to shareholders. Corporate and Wall Street power was counterbalanced by labor unions, farm cooperatives, fair-trade laws allowing small retailers to compete with large chain stores, and government regulation of financial markets and banks to protect small investors. In 1952, economist John Kenneth Galbraith wrote that maintaining a countervailing balance to the power of large corporations and Wall Street was "perhaps the major peacetime function of the federal government."

From 1947 to 1980 increases in real income for working-class households in the United States kept pace with the income increases of the top 5 percent. Corporate social responsibility was evident in this 1981 statement by the Business Roundtable, an organization of U.S. business leaders: "Corporations have a responsibility, first of all, to make available to the public quality goods and services at fair prices, thereby earning a profit that attracts investment to continue and enhance the enterprise, provide jobs, and build the economy. . . . Business and society have a symbiotic relationship: The long-term viability of the corporation depends upon its responsibility to the society of which it is a part. And the well-being of society depends upon profitable and responsible business enterprises."

No company better "walked the walk" than the pharmaceutical

giant Merck. Beginning in 1987, *Fortune* magazine crowned Merck the nation's most admired corporation for an unprecedented seven consecutive years. The CEO, Dr. P. Roy Vagelos, was a highly respected scientist. In 1990 Merck announced it would attempt to keep the price increases of its drugs below the rate of inflation. How quaint, in comparison to current drug company greed: Between 2008 and 2016, brand-name oral and injectable drug prices in the United States increased six and ten times faster, respectively, than inflation.

Under the leadership of Dr. Vagelos, Merck scientists discovered that a single dose of ivermectin (their best-selling antiparasitic veterinary drug) given to people once a year would prevent river blindness, a tropical disease that occurs primarily in sub-Saharan Africa. Merck made the drug available at no charge through a partnership with the World Health Organization.

Anne Case and Nobel laureate Angus Deaton, both professors of economics at Princeton University, describe these postwar years in the United States as the period of "blue collar aristocracy." This shared prosperity was accompanied by better-than-average longevity in the United States compared to other wealthy nations, despite our spending only slightly more on health care.

Of course, the postwar decades were not free of challenges, among them the Cold War, systemic racism, poverty, the Vietnam War, and widespread discrimination based on gender and sexual orientation. There was plenty of room for improvement, but still, most working- and middle-class Americans thrived as they received a generous share of the nation's growing prosperity. And corporations in the United States played a pivotal role.

UNRAVELING OF THE POSTWAR
SOCIAL CONTRACT

The unraveling started slowly with the deregulation of railroads under President Ford in 1976 and the airline industry under President Carter in 1978. But the real turning point came in 1981 when President Reagan, in his first inaugural address, proclaimed: "Government is not the solution to our problem, government is the problem." The guru behind this wisdom, the Nobel Prize–winning economist Milton Friedman, equated laissez-faire capitalism with political freedom. Friedman argued that the role of government should be limited to three functions: "to preserve law and order, to enforce private contracts, [and] to foster competitive markets." He and Alan Greenspan, the Reagan-appointed chair of the Federal Reserve, maintained that allowing markets to regulate themselves would create the greatest prosperity. Other conservative economists concurred, arguing society was best served by corporate managers "single-mindedly working to maximize shareholder value," as measured by the price of shares on the stock market.

President Reagan combined broad deregulation with minimal enforcement of antitrust laws. In 1982 the savings and loan banks that financed local home mortgages were freed to make risky investments that were in large part protected by federal insurance. The number of hostile takeovers of companies worth more than a billion dollars increased more than tenfold from the 1970s to the 1980s.

To preempt such takeovers, corporate management often participated in leveraged buyouts, in which they purchased controlling shares of their own companies. Both takeovers and buyouts required borrowing vast amounts of capital, typically provided by loans in the form of high-yield or "junk" bonds. In any case, corporate executives came under great pressure to squeeze out all the profits possible for shareholders and other investors; if they didn't, they would be replaced by managers who would. Payroll, the largest expense for corporations, became the first target for management. This led to decreasing the number of workers, reducing wages, limiting the role of unions, and outsourcing jobs to decrease labor costs.

These pro–big business and pro–Wall Street changes did not come solely from Republicans. During his first two years in office,

President Clinton supported the adoption of the North American Free Trade Agreement and the creation of the World Trade Organization; Robert Reich, Clinton's secretary of labor, considered both to be of "central importance to big business." NAFTA suppressed American workers' wages, thereby increasing corporate bottom lines. The WTO became the organization that protected the pharmaceutical companies by enforcing U.S. patent regulations globally.* To this day, protection of drug patents and monopoly pricing remain at the core of U.S. trade negotiations.

Publicly held corporations came under pressure to focus on their bottom line, the percentage of workers belonging to unions declined by more than two-thirds (from 35 percent in the mid-1950s to 10.3 percent in 2019), and globalization and automation took American jobs, but the government could not (or would not) provide an effective "countervailing balance" to protect Americans from profit-maximizing corporate behavior. The days when working Americans prospered in proportion to the American economy faded into the history books. And, as we will see, this change in corporate priorities profoundly affected the way that pharmaceutical and medical-technology industries produced and disseminated new medical knowledge.

In 1997 the Business Roundtable unapologetically acknowledged the narrowing of corporate priorities: "The principal objective of

---

* The origin of the WTO can be traced directly back to pharmaceutical company interests. In 1981 Pfizer CEO Edmund Pratt became the chair of the Advisory Committee for Trade Negotiations, a business-sector committee that provided advice on U.S. trade policy. The committee's recommendations were eventually integrated into the 1994 Uruguay Round of Multilateral Trade Negotiations, which culminated in the establishment of the WTO. These negotiations led to the adoption of the Trade-Related Aspects of Intellectual Property Rights (TRIPS), which bound the more than one hundred WTO countries to honor twenty-year patents on new drugs, with some exceptions for middle- and low-income countries. With the TRIPS agreement enforcing patents globally, Big Pharma entered the current enormously profitable era of global marketing.

a business enterprise is to generate economic returns to its owners. . . . If the CEO and the directors are not focused on shareholder
value, it may be less likely the corporation will realize that value."
The measure of corporate performance became the short-term bottom line. Maximum greed, restrained only by the reputational drag
on stock value, was no longer a pejorative. Quite the contrary — it
had become the measure of managerial performance. To align executive and shareholder interests, CEO remuneration was tied to
shareholder value, with stock grants and options accounting for
more than 60 percent of CEO compensation. CEO salaries skyrocketed from "only" 30 times more than that of their average employees in 1976 to 276 times greater in 2017, more than twice as much
as in thirteen other wealthy countries. Lest this be considered old
news, by 2019 the ratio of CEO to average worker compensation
had grown to 320. And then, as ordinary Americans struggled economically during the first thirteen months of the coronavirus pandemic, American billionaires' net worth increased by an astounding
55 percent. The United States had fully entered the era of gloves-off
capitalism.

As part of the pharmaceutical industry's strategy to maximize
profits, since 1998 it has spent almost 50 percent more on lobbying in the United States than any other industry.* In the year and a
half preceding the 2016 election, the pharmaceutical industry spent
an average of $450,000 on lobbying *for each of the 535 members of
Congress.* In 2019, with legislation to control drug prices on the table, the pharmaceutical and health-products industry spent almost
$300 million on lobbying, maintaining three lobbyists for each
member of Congress. As we'll see in chapter 8, for Big Pharma, it
was money very well spent.

The curtain started to descend on Merck's golden era of corporate social responsibility in 1994, when its scientist/CEO Vagelos reached the mandatory retirement age of sixty-five. He was re-

---

* The pharmaceutical industry spent $3,785,353,325, compared to the
$2,622,379,004 spent by the next-highest-spending industry, insurance.

placed by Ray Gilmartin, whose graduate degree was in business, not science. As discussed in chapter 1, ten years after Gilmartin took the helm, Merck was publicly exposed for having failed to act on clear evidence of increased cardiovascular risk associated with Vioxx, but not before the drug killed an estimated tens of thousands of Americans. Merck withdrew Vioxx from the market and Gilmartin opted for early retirement. The *Financial Times* wrote that Merck's "descent into a morass of legal attacks, regulatory scrutiny and ethical doubt must rank as one of the hardest and most sudden corporate falls from grace in recent memory."

But Merck was by no means alone in this behavior. Between 1991 and 2017, drug companies paid more than $38 billion in fines and settlements to federal and state governments; many of those companies — including Merck — pleaded guilty to having misled doctors and the public about drug safety and efficacy. (GlaxoSmithKline led the pack, with $7.9 billion in financial penalties.) Even so, that seemingly enormous amount of money represents a minuscule percentage of drug company profits over the same period, and drug company executives were rarely jailed for wrongdoing. Increasing sales by misleading American doctors (even when this involved breaking the law) had become central to Big Pharma's business strategy.

INCREASED INEQUALITY IS BAD
FOR HEALTH

Since 1980, corporate profits in the United States have grown three and a half times more than they had in the thirty-five years after World War II.* But, in sharp contrast to the egalitarianism of the postwar years, when the least wealthy Americans enjoyed the greatest income growth (by percentage), since 1980, median household

---

* Between 1947 and 1980, annual corporate profits increased by $502 billion. Between 1980 and 2015, they rose by $2,211 billion.

income has increased less than one-tenth as much as the income of the top 1 percent: 22 percent versus 242 percent. The growing inequality has hit African-American households particularly hard: The wealth gap between white and Black U.S. households nearly tripled, from $85,000 in 1984 to $236,500 in 2009. The median wealth of non-retired Black families is now less than 10 percent of white families' ($13,460 versus $142,180). And the median wealth of African-Americans *with* college degrees is less than that of whites *without* college degrees.

The United States now has among the least equal distributions of income and highest poverty rates of the thirty-seven OECD nations. In 2019, the U.S. minimum wage ($7.25 per hour) actually provided 30 percent *less* purchasing power than the country's minimum wage had fifty years earlier. With the rising costs of housing, health care, and education, along with growing indebtedness, people living in the United States are now more than twice as likely to be stressed by rent, mortgage, or the cost of meals as citizens in ten other wealthy countries.

Perhaps the most dramatic evidence of the erosion of our social contract is the *reduction* in the share of total income going to the one-half of U.S. households earning less than the median income compared to the *increase* going to the highest-earning 1 percent. In kitchen table terms, from 1980 to 2015, the amount of income going to households earning less than $55,775 shrank so drastically that it is as if each below-median household is now donating $20,800 ($8,000 per person in an average household of 2.6 people) to the 1 percent earning at least $480,000 per year.* It doesn't take much

---

* In 1980, the bottom 50 percent earned 20 percent of the total income earned by all Americans; by 2015 this had decreased to 12.5 percent. Conversely, the top 1 percent earned 11 percent of total income in 1980, and this had increased to 20 percent by 2015. Based on the $17.5 trillion total American income in 2020:

0.2 (share of total income to bottom 50% in 1980) x $17.5 trillion (total current income) = $3.5 trillion has declined to 0.125 (share

imagination to appreciate the impact of this transfer on the financial burden experienced by the less wealthy half of Americans.

The French economist Thomas Piketty wrote in 2017: "What primarily characterizes the United States at the moment is a record level of inequality of income from labor (*probably higher than in any other society at any time in the past, anywhere in the world*)" (italics mine). Piketty subsequently used the phrase "inequality regime" to highlight the overall context of economic, political, and social power — like that which now exists in the United States as a result of changes that have taken place since 1980 — that maintains the inequality in our society. The small gains in income for non-wealthy Americans are more than consumed by the rising cost of health care. From 2008 to 2018, the increased cost of employees' share of health-insurance premiums and out-of-pocket deductibles exceeded the growth in the median income in every state in the country.

Further exacerbating this imbalance in the distribution of income, the increased cost of health care squeezes funding from virtually everything else in state and local budgets, which — unlike the federal budget — must be balanced each year. In Massachusetts, for example, the 63 percent increase in state health-care spending between 2001 and 2014 has been offset by cuts in public higher education (down by 26 percent), early childhood education and child care (down by 28 percent), local aid (down by 44 percent), and support for parks and recreation (down by 43 percent).

These societal changes have had profound effects on Americans' health and well-being and — as will be shown in further chapters — have prevented our health-care system from responding adequately.

---

of total income to bottom 50% in 2015) x \$17.5 trillion = \$2.19 trillion. Therefore \$1.31 trillion transferred from bottom 50% to top 1%; \$1,310 billion/(329 million Americans/2) = \$8,000 per person in bottom 50% transferred to top 1%, \$8,000 x 2.6/household = \$20,700 transferred up from bottom 50% to top 1%.

The graph reproduced below shows that the greater a country's income inequality, the worse its health and social problems.*

Note how closely all countries except the United States hew to the trend line. Once again, we are an extreme outlier, with by far the highest income inequality and by far the worst rate of health and social problems.

Persistent racial disparities in Americans' health provide part of the explanation. Despite continuing improvement, pre-pandemic longevity for Black, non-Hispanic Americans was still 3.7 years

---

* *Income inequality* is defined as the income ratio of the highest and lowest quintiles in a country. The Index of Health and Social Problems combines data on "life expectancy, mental illness, obesity, infant mortality, teenage births, homicides, imprisonment, educational attainment, distrust and social mobility."

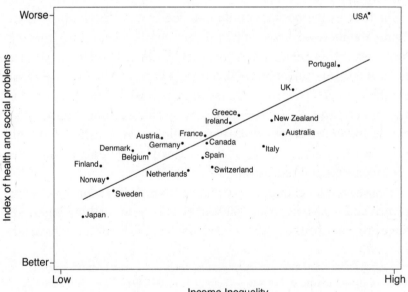

Association between income inequality and health and social problems *Kate Pickett and Richard Wilkinson, "Income Inequality and Health: A Causal Review," Social Science and Medicine 128. Copyright © 2015 by Elsevier Ltd. Reprinted with permission from Elsevier.*

less than for white Americans. And beginning in the late 1990s, another issue became a major factor in the relative decline of Americans' health: The era of "blue collar aristocracy" for white Americans came to a screeching halt. The death rate of white Americans with no more than a high school degree, which had been declining in parallel with that of the rest of Americans, abruptly started to *increase*. Since the early 1990s, that death rate has *grown* by 25 percent. During those same years, the death rate for white Americans with college degrees *decreased* by 45 percent. Today, whites with less than a college education earn only half as much and have four times the annual risk of dying as those who graduated from college.

The increasing death rate for white, non-Hispanic Americans without college degrees has rightfully focused attention on the growing toll of so-called diseases of despair: opioid overdose, alcohol poisoning, suicide, and chronic liver disease. Indeed, these problems are a scourge, increasing, for example, the death rate for people between the ages of fifty and fifty-four in this group by about 100/100,000 (in plain English, one additional death per thousand each year) between 1998 and 2015.

The loss of life due to what Professors Case and Deaton call "deaths of despair" in non-college-educated whites in the United States is devastating, but focusing solely on these deaths obscures the broader health problem: Despair-related causes of death explain only about *one-quarter of the higher death rate in this group.** Increases in routine causes of death — especially cardiovascular disease — play a much greater role, and doctors cannot provide ad-

---

* In 1999 the death rate for white, non-Hispanic Americans ages fifty to fifty-four was 722/100,000. Had this number continued to decline at 1.7 percent per year through 2015, as it had between 1979 and 1999, the death rate would have fallen to approximately $722 - [(0.017 \times 16) \times 722] = 526$. Instead, their death rate increased to 927/100,000, or 401/100,000 more than expected. During those years, deaths due to "diseases of despair" increased by approximately 100/100,000 for white American non–college graduates in this age group (see Case and Deaton, "Mortality and Morbidity," 403, 416, 444).

equate treatment with the tools available to them. The upstream cause is a combination of fewer job opportunities, lower wages, and less social engagement than their parents enjoyed — in general, what Case and Deaton described as "the failure of life to turn out as expected." And this has created an enormous public health issue.

The root of this problem, as we'll see in subsequent chapters, is that the very societal changes responsible for rising corporate profits and increasing maldistribution of wealth in the United States *disempower* our health-care system and prevent us from addressing the consequent health needs. Professors Case and Deaton described this in their book *Deaths of Despair and the Future of Capitalism*: "Warren Buffett likened the effects of healthcare costs on American business to those of a tapeworm; we think of them as a cancer that has metastasized throughout the economy, strangling its ability to deliver what Americans need." And the primary tool used by the industry to impose this cancer is manipulation of the medical "knowledge" it delivers to health-care professionals in order to maximize its profits.

If, as our questioning business student assumed and many experts have claimed, the health of Americans did actually improve as a result of and in proportion to the use of more expensive technology, we could have a spirited debate about whether our unparalleled health-care spending is the best use of our national resources. But the facts show just the opposite: Americans have become less healthy compared to the rest of the developed world since our technology-driven health-care spending spree began.

Victor Fuchs, emeritus professor of economics and health research and policy at Stanford University (crowned the "dean of health economists" by *New York Times* columnist David Leonhardt), agrees that about half the excess U.S. spending on health care is the result of using more expensive services. But unlike Professors Cutler and Gruber, he argues that we don't need these more expensive diagnostic tests and treatments to achieve optimal health. Professor Fuchs wrote in *JAMA*, "It is questionable whether this more expensive mix" of care — including our greater number of specialist and subspecialist visits, our greater number of expensive scans and diagnostic studies, and the more expensive drugs that we use more frequently — "produces better health outcomes." He added that examples of harm in other wealthy countries from reliance on less expensive therapies are "difficult to find."

This leaves us with the question: Why have well-intentioned, hardworking, intellectually disciplined doctors allowed the "knowledge" that guides U.S. health care to be taken over by corporate interests, which have enlisted them as agents of this profit- (rather than health-) maximizing system? But before discussing how drug companies have gained such influence over doctors, we must address the issue of how doctors are taught to receive new information and which sources of information to trust.

6

HOW DOCTORS KNOW

The social acceptance of a knowledge claim always serves
to benefit certain interest groups in the society and to disadvantage others.

— STEVE FULLER, *Social Epistemology*

The previous chapter described American corporations' transition from addressing the needs of a broad array of stakeholders to focusing narrowly on their own financial bottom
line. And coincident with this transition, the cost of our health care
rose rapidly even as our health declined in comparison to that of
the citizens of other wealthy countries.

Currently, nothing is driving up the cost of American health care
as much as our spending on prescription drugs.* In 1980, Americans were spending the same amount per person as were the citizens of other wealthy nations. Now we spend almost twice as much,
and the high price of brand-name drugs in the United States is a significant contributing factor.

---

* The next-three-highest categories are administrative costs, high-cost/
high-volume procedures (like coronary angioplasty, cesarean sections, and
joint replacements), and high-cost imaging such as CT scans and MRIs.

Why do drugmakers charge so much more in the United States than in other countries? Because they can. Unlike other countries, the United States does not regulate drug prices, and it does not have a national program to determine whether new drugs provide added value that justifies their higher cost. This leaves U.S. manufacturers in the commercially advantageous position of holding a monopoly on their brand-name drugs *and* being allowed to set their own prices. Not surprisingly, in 2018 the prices of brand-name prescription drugs in the United States were three and a half times higher than in the other OECD countries.

Conventional wisdom explains our higher drug spending with a simple dictum: "It's the prices, stupid." But the first principle of business is that revenues are not determined by price alone. More than the high *price* of brand-name drugs, it's the *volume* of prescriptions doctors write that really drives up spending. And the primary factor behind that volume is doctors' belief that prescribing these new drugs is in their patients' best interests.

In fact, shaping doctors' beliefs has become the drug companies' primary job. To this end, the majority of the largest drug companies spend twice as much on sales and marketing as they do on research and development of the drugs themselves. And a quarter of those companies spend *ten times as much on sales and marketing*.

But the first step in influencing health-care professionals' beliefs about new drugs is not marketing; rather it is manipulating their primary sources of medical information: the reports of clinical trials and systematic reviews published in respected peer-reviewed journals and the clinical practice guidelines (CPGs) issued by reputable organizations. Doctors have been taught that relying on these sources is in their patients' best interests. But more often than not, the evidence published in peer-reviewed articles in leading medical journals — the "evidence" good doctors rely on — is incomplete, misleading, or both. How have the drug companies gained control of the clinical trial results that doctors use to inform their patient care?

LEARNING THE MEDICAL PARADIGM

To begin to answer this question, I turn to a conversation I had with Dr. Jones (not his real name), a seventyish, soft-spoken, approachable professor who is a practicing physician and an executive at a prestigious medical center. Dr. Jones's primary responsibility as an executive is overseeing the research contracts between his institution and the pharmaceutical companies that sponsor the clinical trials conducted there. Many of his own patients are enrolled in these studies.

We started chatting about the quality of the scientific information available to doctors. I asked if he was aware that when manuscripts reporting the results of corporate-sponsored clinical studies were submitted to medical journals for publication, the underlying data from the studies were kept strictly confidential and remained unavailable to peer reviewers. And further, without access to the underlying data, peer reviewers and journal editors had no way to verify the accuracy and completeness of submitted reports.

No, Dr. Jones hadn't realized that. Then, after he quickly thought through the implications, he lamented, "I feel stupid." If peer reviewers, journal editors, and *even many of the authors* could not independently verify the accuracy and completeness of the data being reported, how could he rely on this unvetted information to guide patient care?

The really scary part of this story is that Dr. Jones's lack of awareness about the absence of data transparency in clinical research was in no way unusual. In my experience, very few doctors know that manufacturers routinely keep their clinical trial data confidential. And even fewer understand the process by which this proprietary information is transformed into the scientific evidence they rely on to inform their clinical decisions.

How does it happen that so many highly intelligent, dedicated physicians fail to recognize the flaws in this system that produces the so-called knowledge that serves as the basis of their patient care? Completion of medical school and specialty training is an enormous achievement, which requires mastering vast amounts of scientific information and participating in years of supervised practice. Even more important than learning the requisite medical science and technical skills, doctors learn how to continue to acquire the knowledge they will need throughout their careers as the practice of medicine evolves. The most significant part of this education is not the specific content; it's learning how to determine which information warrants acceptance as professional knowledge.

Doctors are taught to rely on medical journals, clinical practice guidelines, continuing medical education lectures, and, sometimes, drug company marketing. These sources, they are told, use clinical trial results as their basic building blocks. Moreover, the material found in these sources has undergone stringent peer review with a thorough, independent analysis of the data, ensuring its accuracy, reasonable completeness, and, therefore, trustworthiness.

But unbeknownst to the doctors, the ground has shifted beneath their feet. No longer can they be assured of any of this. An enormous gap exists between doctors' faith in the scientific evidence they receive from trusted sources and the cold hard fact that much of this so-called scientific evidence has reached them because of its commercial value. Its accuracy and completeness have *not* been independently confirmed by the peer-review process, and whether or not effective and efficient practice guidelines are being provided is undetermined.

To understand why this gap is tolerated by doctors, we must turn to Thomas Kuhn, professor of philosophy and the history of science and author of the 1962 surprise bestseller *The Structure of Scientific Revolutions*. Kuhn's most revolutionary insight was that the unspoken truths of "normal science" (meaning the usual practice of scientific inquiry, based on accepted methods) shared by groups of experts do not necessarily reflect the most accurate or relevant *scientifically* revealed knowledge. Rather, these "truths" are first and foremost *sociologically* determined, emerging from the group's implicitly shared rules defining the boundaries of legitimate inquiry for their particular field. Kuhn used what has subsequently become the clichéd concept of a paradigm to define the shared beliefs that lie at the core of scientific disciplines, and he showed that adherence to these beliefs is required for an individual to be accepted into a community of scientific experts (including, for our purposes, physicians).

Think of the classic example of Galileo, who was "vehemently suspected of heresy" in 1633 by the Catholic Church. His crime was publishing a book that — based on scientific observations made with a new technology, the telescope — argued against the accepted dogma that Earth was the center of the universe. Galileo had to publicly recant his "errors and heresies," his book was banned, and he was sentenced to house arrest for the remainder of his life. Like Galileo's geocentric-believing colleagues, today's well-meaning physicians are in the unfortunate position of having to base much of their practice on similarly dysfunctional dogma.

In 1981, during my Robert Wood Johnson fellowship in family medicine, I studied Kuhn. At the time, the paradigm that was of primary concern to academic family physicians was the biomedical model that defined disease as the result of machinelike malfunction at the molecular, cellular, or physiological level. Based on this paradigm, the appropriate tools to restore and maintain health were narrowed almost exclusively to pharmaceuticals and medical procedures.

In the academic milieu of the university medical center, the biomedical paradigm drowned out family medicine's broader holistic vision that health and disease are also — and, in fact, more likely to be — the result of interaction between individuals and their social, cultural, and physical environment. Indeed, as mentioned earlier, only about 20 percent of a population's health is determined by medical care; most of the rest is determined by these external factors.

Our group of fellows spent many hours in seminars critiquing influential articles published in respected medical journals. We addressed problems with research design, statistical analysis, and conclusions, as well as the artificial constraints imposed by the bio-medical model. Although there was plenty to criticize, the faults were not — as far as we could discern — the result of commercial bias.

Only in retrospect did I realize that my fellowship had come at a watershed moment in American medicine. The paradigm of biomedicine was about to be subsumed by the paradigm of evidence-based medicine (EBM), which called for doctors to base their decisions on scientific evidence from high-quality clinical trials published in peer-reviewed medical journals. But at the same time, the funding and then the control of clinical trial design, conduct, analysis, and reporting were being taken over by the drug companies. As this transition progressed, growing commercial influence ensured that the primary purpose of most new knowledge became supporting the use of new drugs and other biomedical interventions. American health care quietly and almost invisibly slipped into the current era of commercially based medicine.

THE COMMERCIALIZATION OF
CLINICAL RESEARCH

In the mid-1960s, more than half the grant applications to the National Institutes of Health for support of medical research were successfully funded, which allowed academic researchers the luxury of remaining independent of drug company influence. By the late 1970s, however, the success rate of grant applications had fallen to about one-third. At the same time, Senator Ted Kennedy's efforts to increase federal funding for drug development and create a national center for clinical pharmacology were defeated by opposition from the drug industry and medical leaders. Their refusal to participate in a coalition that yielded authority to the federal government foreshadowed the exclusion of government from oversight of pharmaceutical policy that would lead to many of our current problems. (This was consistent with the reduction in gov-

ernment regulation discussed in the previous chapter.) And starting in 1981, President Reagan reduced government support of university-based medical research, pushing academic researchers who
wanted to use their skills and talents into the open arms of industry,
especially the pharmaceutical industry.

Nonprofit research institutions were further drawn into the
commercial realm by the passage of the University and Small Business Patent Procedures Act of 1980, aka the Bayh-Dole Act. This
legislation allowed nonprofit research institutions to commercial-
ize discoveries made by their scientists while conducting federally
funded research and to retain any profits derived from those discoveries. Not surprisingly, lobbying by the institutions that stood
to profit financially from this federal giveaway — universities, academic researchers, and the drug and device companies — had supported the legislation.

As the opportunity for commercial gain from scientific research
permeated universities, Harvard president Derek Bok described
"an uneasy sense that programs to exploit technological development are likely to confuse the university's central commitment to
the pursuit of knowledge and learning by introducing into the very
heart of the academic enterprise a new and powerful motive — the
search for utility and commercial gain." A 1982 article in the journal *Science* summed up Bok's "uneasy sense" as "concern about
whether the academy is selling its soul" to industry. The article
continued, "Scientists who 10 years ago would have snubbed their
academic noses at industrial money now eagerly seek it out."

By 1985, half of the biotechnology research faculty members at
top universities were consultants to industry, and a quarter were
leading commercial research studies. In the mid-1990s, more than a
third of all clinical faculty members at the fifty top American university hospitals were receiving industry money, likely influencing (or
at least consistent with) their choice of therapy for their patients.
Now, if you could put on truth goggles and observe the prestigious
academic physicians striding down the halls of university hospitals
surrounded by their bevy of trainees, you might see, instead of their

clean white doctor coats, coats plastered with prescription-drug ads and drug company logos, making them look more like NASCAR drivers than accomplished medical professionals.

During the 1990s control of clinical research underwent a major transition. In 1991, academic medical centers (AMCs) received 80 percent of the money that industry spent to fund clinical trials. The drug companies relied on AMCs for expertise in designing studies and enrolling patients. But the academics lost this influence over clinical trials when the pharmaceutical industry developed in-house research design capacity and turned to independent for-profit medical research companies (contract research organizations, or CROs) to conduct the research. This avoided university red tape and overhead and, most important, transferred much of the oversight of the research from academia to industry. Under this new arrangement, drug company sponsors could design the studies, analyze the data, and prioritize those results that would maximize the financial return on their research investment.

From a business point of view, there was nothing unethical about the drug companies' turn to CROs. But from the perspective of preserving the integrity and relevance of the clinical research findings that doctors rely on, increased commercial control neutralized the ability of academic researchers to act as "disinterested" guardians of scientific integrity. Academic researchers became part of the drug company–funded team whose primary mission was to produce study results that would enhance drug sales.

In 1997, Dr. Drummond Rennie, deputy editor of the *Journal of the American Medical Association*, wrote that competition for industry funding of research would cause AMCs to engage in "a race to the ethical bottom." That's worth a second look: The deputy editor of one of the most prestigious medical journals in the world was observing that the nation's most prestigious academic medical centers were willing to compromise the autonomy of their research — the basis of doctors' treatment decisions — to maintain funding from commercial sources.

Then it got worse. Can you guess who started buying up the

CROs? Most people answer, "The drug companies." Wrong. It was big advertising agencies (Omnicom, Interpublic, and WPP, among others). Why would advertising agencies buy research companies? As reported in the *New York Times:* "Their intention is to work side by side with scientists, directing research toward drugs the marketers think could be best sellers." This is unadulterated Orwell — advertising agencies owning research companies whose findings they would be hired to spin, hype, and broadcast for maximum corporate advantage.

Not everyone was sitting idly by while the integrity of American research was under attack. In September 2001, the editors of twelve of the world's most influential medical journals (including *NEJM, JAMA,* and *The Lancet*) published an unprecedented joint statement warning that the validity of clinical trial reports was threatened by commercial influence, and the use of the studies for marketing was making "a mockery of clinical investigation." The journal editors did not pull any punches, noting that, as things stood, "investigators may have little or no input into trial design, no access to the raw data, and limited participation in data interpretation. These terms are draconian for self-respecting scientists, but many have accepted them because they know that if they do not, the sponsor will find someone else who will." The editors concluded that these arrangements were a violation of academic independence that "not only erode the fabric of intellectual inquiry that has fostered so much high-quality clinical research, but also make medical journals party to potential misrepresentation."

But the editors' clarion call went virtually unnoticed, in part because the integrity of medical knowledge was not the only thing under attack that week. The journal editors' alarm was completely overshadowed by the tragic events of September 11, 2001.

By 2004, the percentage of drug company–funded clinical research performed by AMCs had plummeted from 80 to 26 percent (the rest going to CROs). In that year, a survey of research contracts between AMCs and industry funders documented the drug companies' growing control of clinical research. The results showed that

80 percent of these contracts let the commercial funders "own the data produced by the research," and 50 percent granted the sponsor the right to "write up the results for publication and *the investigators may review the manuscript and suggest revisions*" (italics mine). In plain English, this means that half the research contracts between drug companies and academic institutions — the partnerships with the best opportunity to uphold the standards of independent science — allowed the commercial sponsors to ghostwrite the manuscripts and relegated the named authors to the position of "suggesting" revisions. In return for the prestige and career advancement associated with participation in commercially sponsored clinical trials, academic researchers were signing away their names and relinquishing control over what was being attributed to them.

The transition continued. Between 2006 and 2014, the percentage of clinical trials funded by industry increased from 77 to 86 percent while those funded by the NIH declined from 23 to 14 percent. Funding an even higher percentage of clinical trials gave industry even more control over the design of the studies and the data produced.

The Problems with Drug Company Control of Research

Commercial sponsors can bias the design of clinical trials to favor their drugs in several ways, some of them quite subtle. For one, the people enrolled in a study may not reflect the people to whom the study results will be applied in terms of age, sex, race, disease, or medication history. Merck used this tactic to bias the VIGOR study of Vioxx discussed in chapter 1; as a result of Merck's criteria for participation in the study, more than half the subjects were taking steroids (such as prednisone) for arthritis, even though steroids are rarely taken chronically by the people to whom the results would be applied. When the underlying data for the VIGOR study were finally made available during one of the last Vioxx lawsuits, we found that among the people not taking steroids concurrently with Vioxx, there was absolutely no reduction in serious gastrointestinal problems associated with Vioxx compared to naproxen. In other words, the whole justification for patients to take Vioxx — that it had a gastrointestinal safety benefit for most people — had been manufactured by the study design; in reality, only the small percentage of people taking steroids concurrently with other anti-inflammatory medication benefited. This means that the vast majority of the up to sixty thousand Americans who died as a result of taking Vioxx could have received the same anti-inflammatory effect from far safer and less expensive over-the-counter Advil or Aleve.

Another way to bias study results is to have "run-in" phases that screen potential study participants to make sure they will take the medicine as prescribed and not experience side effects. This was done in the huge Heart Protection Study evaluating cholesterol-lowering simvastatin (Zocor); more than a third of the people who entered the run-in phase were ultimately excluded from participation in the study. The benefits of statin therapy found in this "enriched" study population were then applied to all potential candidates for the drug, not just those resembling the ideal patients who had made it through the run-in period.

Manufacturers can also bias the study by administering drugs in doses that may not be the most effective or may be higher than needed or approved. Pfizer used this tactic in the forced-titration study of Neurontin discussed in chapter 2, producing a high rate of side effects that functionally unblinded the study.

In addition, the sponsor's drug is routinely *not compared* to the best available therapy, so positive results from one or even many clinical trials do not necessarily provide evidence of clinical superiority. This is the case with Humira, which has been by far the highest revenue-generating drug in the United States and the most heavily advertised for treating rheumatoid arthritis (RA). The current FDA-approved label reports the results of five of the manufacturer-sponsored clinical trials of Humira for the treatment of RA. Four of the studies simply showed that Humira was more effective than no treatment. But this study design does not reflect the real-world clinical practice. The choice for a patient requiring new or additional therapy is not to start Humira or do nothing. The choice is whether to start Humira or use the best available conventional therapy, so comparing Humira to nothing does not produce clinically useful evidence. The fifth trial was more informative: It compared Humira to a conventional drug therapy, methotrexate. The study found that Humira (the discounted retail price of which is now about $72,000 per year) was no more effective than methotrexate (which costs about $480 per year). Just to be clear, the manufacturer's study results show that, as single-drug therapy, Humira is no more effective — *but costs 150 times more* — than methotrexate.*

But *none* of the clinical trials tested the combination of Humira plus methotrexate against the most effective conventional therapy for RA, which is triple therapy with three conventional anti-inflammatory drugs — methotrexate, hydroxychloroquine, and sulfasalazine. Although there remains no clinical evidence that Humira (or the similar Enbrel or Remicade) plus methotrexate is superior to triple therapy for the treatment of RA, the combination is used forty times more frequently than triple therapy when a single conventional drug has failed to provide adequate relief. This is despite the fact that triple therapy with conventional drugs costs less than 2 percent of what Humira plus methotrexate costs.

Finally, when lifestyle changes — like maintaining a healthy diet, exercising, and stopping smoking — potentially have a favorable impact on conditions such as heart attack, stroke, or diabetes, such changes should be included in clinical trials as a comparator to whatever drug is being tested. This was done in the study of diabetes prevention in people with prediabetes described in chapter 4, producing dramatic results showing lifestyle modification is more

---

* Although patients treated with methotrexate had slightly (non-significantly) better clinical response rates, those treated with Humira showed significantly less progression of joint damage on x-rays.

effective than drug intervention. But lifestyle modification versus statin therapy to prevent cardiovascular disease has not been tested in any major study of statins. Thus, the "knowledge" produced by clinical trials about preventing heart attacks and strokes is limited primarily to knowledge about drug therapy, with information about the benefits of lifestyle modification relegated to less expensive and less reliable observational studies.

Clinical Trial Data Belong to Drug Company Sponsors

The slide shown below, from a Pfizer internal document (presented in litigation involving Pfizer's antidepressant drug Zoloft), reveals in the starkest of terms that the contract language claiming ownership of the data from research funded by commercial sponsors is not just boilerplate.

The second bulleted point unabashedly states that the purpose of Pfizer's clinical trials is to create data "to support, directly or indirectly, marketing of our product." There you have it: The pur-

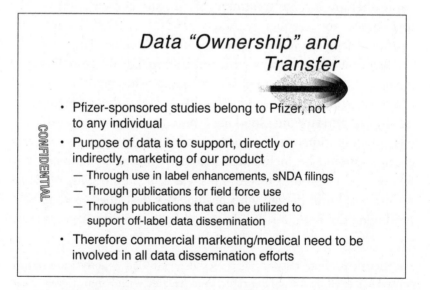

Pfizer's view of data ownership

pose is *not* to support physicians' efforts to optimize the health and well-being of those affected by or at risk of maladies related to the drug in question. No, the purpose of Pfizer's research — and the research of all the major drug companies — is to produce scientific evidence that will enhance their marketing efforts.

This corporate goal is served by limiting the role of academic researchers (as the journal editors' statement of 2001 described) and limiting the ability of peer reviewers to independently verify study findings. Corporate ownership of data allows the drug companies to control the information they make available to physicians and the public.

WHENCE COMES SCIENTIFIC EVIDENCE?

Let's assume Pfizer was telling the truth about its motive for doing clinical studies. Such studies would be designed to maximize the likelihood of producing data that emphasize the benefits and minimize the risks of the company's drugs. But, you may be thinking, science is objective, so what difference does it make who the sponsor is? Doesn't the rigor of the scientific method ensure that, except in rare instances of fraud, research results have been honestly and completely presented?* The answer: not necessarily. Understanding how this works requires a more granular look at the process by which the mountains of data that are collected in each clinical trial are winnowed down to the brief results and conclusions presented as scientific evidence in the medical journals that doctors read.

The diagram on page 118 depicts the totality of evidence generated in a clinical trial as dispersed throughout an iceberg. The ob-

---

* In this case, the study would be said to have "internal validity." Assuming accurate analysis, the results could be applied to the chosen study design and population. If the chosen details created a study or included a population that did not represent the people most likely to take the drug in the real world, the study would then be said to lack external validity.

server in the boat with the spyglass represents a doctor searching for the scientific evidence that will inform his or her clinical decision-making. But that observer can see only the small fraction of information from each study that emerges above the water — that is, the results of the trial published in a peer-reviewed medical journal, included in a systematic review or clinical practice guideline, or used for marketing.

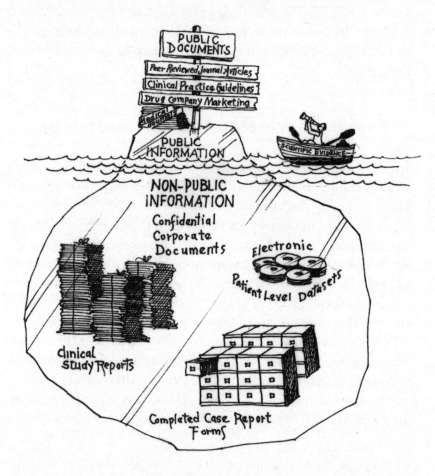

A doctor searching for scientific evidence in clinical trial documents *Illustration by Darby E. Mumford,* adapted from: Peter Doshi et al., "Restoring Invisible and Abandoned Trials: A Call for People to Publish the Findings," *BMJ* (2013):346f2865.

The only way for the doctor to be sure that the scientific evidence visible above the water reasonably represents the voluminous data below the waterline is to have—or, more realistically, for trusted independent experts to have—unfettered access to all the data. So first, let's see what's down there, and then we'll look at who is and is not allowed to see it.

Drug company employees, with the assistance of CROs and in conjunction with academic researchers (usually paid by the drug company), work together to design, conduct, analyze, and write up the study's findings for publication. (Only rarely these days are drug company–sponsored clinical trials conducted independently by academic researchers.) During the study, patients are evaluated initially and in follow-up visits according to the study protocol (study protocols and analysis plans from clinical study reports are often available to the public). This information is recorded on case-report forms that include all data collected at each visit for each participant in the study; data are later de-identified so collected information cannot be attributed to a specific person. These data are transferred to electronic patient-level data sets, which are then analyzed according to the statistical analysis plans. The sponsor or the CRO then compiles a clinical study report (CSR) that includes the study protocol, statistical analysis plan, amendments, and all relevant data in tables and figures. These reports can be several thousand pages long.

Many steps are required to ensure the integrity of this process, but, in addition to accurately entering data on the case-report forms, two are of fundamental importance. The first is correct analysis of the voluminous data collected and stored "below the waterline" to determine the results for the outcome measures that have been specified in the study protocol and statistical analysis plan (as well as unexpected but clinically relevant safety issues that may arise). The second is the accurate and complete reporting of these outcome measures in the published articles that emerge "above the waterline." Most of what practicing doctors believe they know about optimal care comes directly or indirectly from the medical journal articles that report clinical trial results. As a corollary, they've been taught to trust that what they read there has been properly vetted and accurately reflects the relevant raw data hidden below the waterline. Little do they know . . .

Listen to what Dr. Jones had to say when I asked him how things were done at his institution, which, as I've said, is a major medical center where many clinical trials are conducted. According to Dr. Jones, research nurses working for the drug company or CRO would come to his hospital carrying suitcases filled with forms for the patients participating in the clinical trials they were working on at the time. The nurses would transfer data by hand from the hospital's electronic medical records to those case-report forms and leave at the end of the day with the manually updated information in their suitcases. As my discussion with Dr. Jones unfolded, he connected the dots: The nurses were not electronically transferring data from the hospital's computer system to their own. They were not even using computers. They were working with hard copies. There was no easy data sharing. No trail. No simple validation.

Dr. Jones explained that his own research interest was in tracking the outcomes that matter most to patients who receive care at his institution — primarily comparing the quality of life for patients who did and did not receive the study drug. Yet the research sponsors were unwilling to share with him even this basic data for other study sites beyond his hospital. This made it difficult or even impossible to evaluate study-wide data and determine the effects on quality of life.

A survey of articles reporting clinical trial results published in the seven most influential medical journals showed that Dr. Jones's experience was the rule, not the exception. Only 40 percent of the studies had included academic authors (that is, authors not employed by the sponsoring drug company) in data analysis, and only 39 percent of the trials allowed all authors access to the underlying data. This means that most of the academic authors who participate in drug company–sponsored clinical research cannot even check the accuracy of the results for their own studies. The sponsor retains control of this information.

I asked Dr. Jones why the doctors at his hospital didn't drive a harder bargain with the research sponsors. For example, why didn't they demand that sponsors share patient-outcome data from all study sites, rather than limiting them to only the data they could gather from the patients in their own hospital? He said the doctor-researchers were reluctant to push harder in negotiations because they didn't want to risk being excluded from the intellectual, reputational, and financial rewards of participation in clinical research. Recall *JAMA* deputy editor Drummond Rennie describing the weakened negotiating position of AMCs competing with contract research organizations as "a race to the ethical bottom." More than twenty years later, Dr. Jones was caught in that same race.

THE BLIND SPOT IN EVIDENCE-BASED MEDICINE

This "modern" era of acceptance of what appear to be methodologically rigorous reports of clinical trials published in peer-reviewed journals was codified in 1992 as the basic precept of evidence-based medicine (EBM). At the time, this was hailed as a "paradigm shift" in the way doctors determined how to provide the best care. The EBM movement called for replacing "traditional scientific authority" — "because I said so" authority — with doctors learning the skills of searching and critically evaluating relevant medical journal articles to determine optimal patient care for themselves.

In 2004, the EBM movement introduced a more nuanced hierarchy of evidence, the "Grading of Recommendations Assessment, Development, and Evaluation." The GRADE system refined optimal decision-making by evaluating the direction and strength of recommendations based on four factors:

- certainty (or quality) of the evidence
- size of benefits and harms
- cost, acceptability, and health equity
- patient values and preferences

In 2017, the EBM movement was again updated. Recognizing that doctors are limited in both the time and skill required to critically evaluate published research reports, the leaders of the EBM movement called for clinicians to rely on high-quality "processed" overviews of published clinical trial results, especially "trustworthy clinical practice guidelines," as determined by the GRADE framework.*

Over the past thirty years, the EBM movement has played a key role in the founding of the Cochrane Collaboration (responsible for Cochrane Reviews), establishing standards for research, providing benchmarks for developing clinical practice guidelines, and building the knowledge base that doctors turn to for up-to-date information about optimal patient care. But — and this is an enormous *but* — without data transparency and independent analysis, the tools of EBM function as a double-edged sword. When applied to publications reporting the results of commercially funded clinical research, the source of most of the scientific evidence doctors rely on, the techniques of EBM mask the lack of transparency of the data that remain below the waterline.

Looked at another way, when the discipline of EBM is combined with the falsely empowering belief that the skills of EBM allow full evaluation of published research, doctors are placed in much the same predicament as the prisoners in Plato's allegory — chained in a dark cave, mistaking the shadows cast on the cave wall in front of them for reality. Well-meaning doctors mistakenly assume that the

---

* Determination of the credibility of evidence includes evaluation of "study design, risk of bias (study strengths and limitations), precision, consistency (variability in results between studies), directness (applicability), publication bias, magnitude of effect, and dose-response gradients."

reports of clinical studies in peer-reviewed medical journals are accurate and complete distillations of study data; they are unaware that the published results are incompletely vetted, often commercially biased selections from the totality of data. Doctors try their best to practice high-quality medicine, but they don't know what they don't know. Succinctly stated, "Evidence without data is not evidence."

In sum, the principles and skills of EBM are necessary for doctors to make optimal practice decisions, *but they are not sufficient.* And, ironically, this lack of sufficiency is obscured by the truly valuable tools and skills EBM does teach. Rather than providing doctors with the skills to liberate themselves from the dictates of medical authority, EBM drives adherents into the less apparent authority of commercial sponsors. The result is medical care designed to maximize Big Pharma's profits, accompanied by the steadily declining overall health of Americans in comparison to citizens of other wealthy nation and punctuated by occasional dramatic therapeutic success.

EXAMINING THE SOURCES

The information doctors rely upon to fulfill their role as learned intermediaries — interpreters of the scientific evidence of the benefits and harms of therapeutic options that are then shared with individual patients — comes primarily from three sources: peer-reviewed journal reports of clinical trials, review articles and meta-analyses, and clinical practice guidelines. This section examines how adequately each of these sources meets the requirements of good science. And remember: The point here is not that all the information presented in them is inaccurate or incomplete. The problem is that distinguishing between accurate and complete versus inaccurate or incomplete information is impossible without real-time access to the data hidden below the waterline.

1. Reports of Randomized Controlled Clinical Trials in Peer-Reviewed Medical Journals

EBM teaches doctors that the findings of methodologically rigorous randomized controlled trials published in peer-reviewed medical journals provide the scientific foundation for the practice of medicine. But today, six out of seven clinical trials are commercially funded, and this matters for several reasons.

First, commercially sponsored clinical trials report favorable conclusions about the sponsor's drug or device 34 percent more frequently than non-commercially funded trials do. Moreover, when the principal investigator of a published clinical trial has financial ties to the manufacturer of the drug being studied — as is the case in two-thirds of published clinical trials — the odds are three and a half times greater that the study will report that the drug is effective. And amplifying the significant bias found in commercially sponsored studies, 97 percent of the studies most frequently cited in the literature are funded by industry. This means that more than nineteen out of twenty of the studies having the greatest impact on doctors' practice are commercially funded — and significantly more likely to show positive results for the sponsor's product.

Next, researchers from the German Institute for Quality and Efficiency in Health Care provide troubling evidence for those who assume that clinical trial results published in peer-reviewed journals are reasonably complete. The study focused on the reporting of outcomes that are most important to patients — effects on survival, relief of symptoms, quality of life, and side effects. They found that only 23 percent of these four patient-relevant outcomes had been reported in the 101 medical journal articles for which they had received below-the-waterline data.

Another study showed only 40 percent of the primary outcomes reported in medical journals were consistent with the outcomes that had been declared before the trial began, meaning the primary outcomes specified in online registries. (Full disclosure: I served as a peer reviewer for this study.) What's wrong with switching a study's primary end point after the study is completed as long as the data are accurate? It's like aiming an arrow at a target, hitting something else, then claiming that that's what you'd been aiming for all along. For sure, sometimes studies produce unexpected and potentially important results. These are called post hoc findings. It's perfectly reasonable to report them, as long as it's made clear they weren't the originally specified primary outcomes. If such a finding has potential clinical importance, a new study aiming at this target can be done. But reporting a miss as if it were a hit, by altering the primary outcome after a study has been completed, is a flagrant methodological foul.

These obvious biases in trusted peer-reviewed reports of clinical trials could be prevented with ease, and the fact that they aren't provides a window into the systematic tolerance of commercial influence on doctors' medical knowledge. The German Institute for Quality and Efficiency in Health Care compared the adequacy of reporting "sufficient information for trial evaluation" in clinical study reports (recall the several-thousand-page CSRs that were depicted as remaining underwater in the diagram on page 118) to the data published in medical journals. The study found that CSRs provided 90 percent of the information on study methods and results that were necessary to evaluate clinical trial findings. In comparison, reports published in medical journals contained only about 50 percent of the necessary information.

This study did not identify exactly which information was missing from the published articles, but it's a safe bet that commercial sponsors did not selectively omit information that favored their products. So why would medical journals not require submission of manufacturers' CSRs, which are already prepared (and, in the rare instances when patient identification or trade secrets were at risk, easily redacted), along with the manuscripts reporting the findings of clinical trials? This would be a quick and inexpensive way to allow peer reviewers and editors to perform a far more thorough evaluation of the accuracy and completeness of the submitted reports *before* publication. This issue is addressed in chapter 8, but the short answer is that it would not be in the journals' financial interests.

2. Review Articles and Meta-Analyses

EBM proponents consider systematic reviews and meta-analyses, which combine data from multiple clinical trials, to be an even more important source of information for health-care practitioners than individual, high-quality, peer-reviewed articles. The number of these overviews being published has increased by 2,500 percent since 1991, and they have become the most frequently cited type of article. But we've already seen how commercial control of the data prevents independent analysis of articles reporting individual clinical trials, and drug company involvement in these compilations of study results adds another opportunity for industry to bias the information available to doctors.

For example, a study of 185 meta-analyses of antidepressants showed that four out of five were sponsored by industry or had authors with financial ties to industry, and almost one-third had at least one author who was employed by the manufacturer of the drug being reviewed. Another study found that meta-analyses with industry funding recommended the drug being studied "without reservations" almost twice as often as studies without industry funding did, 40 percent and 22 percent, respectively.

Worse yet, according to Professor John Ioannidis at Stanford University, a leading authority on systematic reviews and evidence, just 3 percent of all published meta-analyses are methodologically "decent and clinically useful." Professor Ioannidis concluded: "Currently, there is massive production of unnecessary, misleading, and conflicted systematic reviews and meta-analyses. Instead of promoting evidence-based medicine and health care, these instruments often serve mostly as easily produced publishable units or marketing tools."

The most methodologically sound of these reviews, according to EBM proponents, are performed by the Cochrane Collaboration. Cochrane is a global nonprofit organization whose first goal is "to produce high-quality, relevant, up-to-date systematic reviews and other synthesized research evidence to inform health decision-making." Nevertheless, as respected as Cochrane Reviews are, they are not immune to the bias and power of the drug companies.

A brief history of the Cochrane Review of the antiviral drug Tamiflu, used to treat and prevent influenza, demonstrates the difficulty of incorporating complete evidence into systematic — even Cochrane — reviews. A meta-analysis published in 2003 showed Tamiflu reduced the risk of bronchitis, pneumonia, and hospitalizations when given to adults with acute flu-like symptoms. Relying on these findings, a 2006 Cochrane Review concluded that Tamiflu, "in a serious epidemic or pandemic, should be used with other public health measures." The U.S. Department of Defense purchased and stockpiled forty million doses, and the World Health Organization labeled Tamiflu an essential or "core" medication.

Three years later, in the midst of the 2009 swine-flu pandemic, a Japanese pediatrician, Dr. Keiji Hayashi, questioned the Cochrane Review's conclusion. Dr. Hayashi had noticed eight of the ten studies in the 2003 meta-analysis — all sponsored by the drug's manufacturer, Roche — had never been published and the results could not be verified, not even to the limited extent that peer review can verify results. The authors of the 2003 meta-analysis had had access to those unpublished results because four of the six were Roche employees.

Faced with this new information Dr. Tom Jefferson, lead author of the 2006 Cochrane Review, wrote to Roche requesting full CSRs for all ten trials. Roche sent ten- to seventeen-page excerpts from each, a response Dr. Jefferson deemed grossly inadequate and evasive. Much finger-pointing and stonewalling ensued. Finally, after four years of requests from the Cochrane reviewers and the *British Medical Journal,* in 2013 Roche provided the full CSRs for all its seventy-seven trials of Tamiflu.

Based on the additional information sent by Roche, Jefferson and his colleagues updated the Tamiflu review in 2014. The update showed the totality of Tamiflu's benefit was minimal, limited to reducing the duration of flu symptoms in adults from 7 to 6.3 days, without preventing complications, hospitalizations, or deaths. On the negative side, Tamiflu increased the frequency of nausea, vomiting, headaches, and psychiatric symptoms in a significant proportion of those treated. By this time, the United States had stockpiled enough Tamiflu to treat sixty-five million people, ultimately adding $1.3 billion to Roche's bottom line.

In 2017, based in part on the updated Cochrane Review, the World Health Organization downgraded Tamiflu from a "core" drug to a "complementary" drug, a category for drugs that are less cost effective. However, for more than a decade, not just physicians but the U.S. Centers for Disease Control and Prevention, the European Medicines Agency, and the World Health Organization had been making recommendations based on selectively incomplete information. Globally, more than $18 billion was spent on Tamiflu, half of which was purchased by governments for stockpiling. And to be clear, governments don't stockpile medicines that do nothing but reduce the duration of symptoms by two-thirds of a day. They stockpile drugs that can be used when serious public health concerns are at stake.

These Cochrane reviewers spent years obtaining confidential CSRs from the manufacturer and then thousands of person-hours to uncover the truth about Tamiflu. But we cannot depend on efforts as heroic as this group's to get effective reviews; one study showed that 87 percent of Cochrane Review authors and editors had not requested FDA data or other regulatory data and simply relied on publicly available sources (information above the waterline). The reasons for not seeking out unpublished and otherwise unavailable data included — not unexpectedly — limited time and resources on the part of reviewers. In this contest between drugmakers and health-care professionals, the drugmakers clearly have the advantage.

3. Clinical Practice Guidelines

Physicians face increasing time pressure and relentless information overload, so clinical practice guidelines would seem to be the perfect tool to guide optimal patient care. As already mentioned, CPGs tailored to individual patients have become the primary focus of the evidence-based medicine movement. However, although the importance of CPGs is widely recognized, their potential to improve patient care has been severely compromised by ongoing financial ties between guideline writers and the drug companies, as well as the writers' lack of access to full clinical trial data.

A bit of history: In 1989, the Agency for Health Care Policy and Research (AHCPR) was created, in part because of concern at the federal level about "ceaselessly escalating healthcare costs, wide variations in medical practice, [and] evidence that some health services are of little or no value." The agency then contracted with the Institute of Medicine (subsequently renamed the National Academy of Medicine) to help oversee the quality of CPGs. The following year the IOM issued a report that included recommendations to improve the validity and clinical relevance of CPGs and address cost control and comparison of alternative therapies. Most subsequent guidelines, however, failed to comply with the IOM's recommendations. The title of a 1999 study was "Are Guidelines Following Guidelines?" The answer: not so much. Published guidelines met less than half (43 percent) of the IOM standards.

In 2011, the IOM updated its standards for guidelines. Because practice guidelines have such great influence over what doctors prescribe, one of the IOM report's core recommendations called for minimizing financial ties with relevant drugmakers; "whenever possible," committees should have no members with conflicts of interest and certainly no more than a minority of any panel. Moreover, chairs and cochairs of guideline panels should be free of conflicts.

The following year a study found that guidelines on the website of the National Guideline Clearinghouse for the Agency for Healthcare Research and Quality (AHRQ, formerly AHCPR) "demonstrated poor compliance" with those IOM recommendations. The study found 71 percent of chairs and 90 percent of cochairs had relevant financial ties, and compliance with IOM standards had not improved over twenty years. The reason for the IOM's concern about financial ties is provided by a study of ninety-one cancer treatment guidelines. When one or more authors had a financial tie to a particular drug, the guidelines were *seven times* more likely to recommend its use.

In 2017, the AHRQ took a giant step toward bringing guidelines into compliance with the IOM recommendations by requiring its website to show "the extent to which clinical practice guidelines (CPGs) submitted adhere to the standards for trustworthiness." The government's National Guideline Clearinghouse quickly became the go-to site for guidelines and summaries, attracting two hundred thousand visitors per month.

Then it suddenly went dark.

On June 18, 2018, this notice appeared on its website: "The AHRQ National Guideline Clearinghouse (NGC, guideline.gov) Web site will not be available after July 16, 2018, because federal funding through AHRQ will no longer be available to support the NGC as of that date." Due to federal budget cuts, AHRQ defunded it. Canceled. Nixed. Gone. The 2017 document about the trustworthiness of guidelines is now available on the internet only through the Wayback Machine.

Had the clearinghouse website not been defunded, it would make for interesting reading. Two studies examining financial conflicts in guidelines were published in October 2018. One showed that, among eighteen guidelines pertaining to ten "high-revenue" drugs, more than half (57 percent) of guideline authors had conflicts, more than a quarter did not declare them, and fourteen of the eighteen chairs had financial ties to the industry. The other article reported that more than half of gastroenterology guideline authors had financial ties to industry, but only 34 percent had declared them.

The accompanying editorial in *JAMA* asked: "If the specialty societies and other organizations that prepare clinical practice guidelines are unwilling or unable to improve their performance in disclosing, managing, and eliminating financial conflicts of interest, what can be done?" This question takes on much greater importance now that the National Guideline Clearinghouse, which was charged with overseeing compliance with standards that make CPGs trustworthy, has been defunded.

A clue to the recalcitrance of the problem appears in the American College of Rheumatology's statement of principles: "Pharmaceutical and biotechnology companies represent an important source of financial support for the American College of Rheumatology." Perhaps financial relationships played a role when the American Diabetes Association and the American Association of Clinical Endocrinologists recommended the eleven-times-more-expensive insulin analogs over recombinant human insulin for the treatment of type 2 diabetes (chapter 4). Or when the American College of Cardiology and the American Heart Association recommended cholesterol-lowering statin therapy for so many healthy Americans without heart disease (chapter 3).

Doctors' professional societies are dependent on money from industry, even as *JAMA* asks them to eliminate conflicts of interest in their development of guidelines. As the political writer and muckraker Upton Sinclair wrote: "It is difficult to get a man to understand something, when his salary depends on his not understanding it."

The National Academy of Medicine and the U.S. Agency for Healthcare Research and Quality have made repeated appeals, asking for CPGs that are independent of commercial influence, but those appeals have failed. Doctors are left in the untenable position of having the quality of their practice judged by their adherence to guidelines written by experts who often have financial ties to the drug industry and who rarely, if ever, have access to the underlying data necessary to independently evaluate the evidence from the primarily commercially funded clinical trials. And Americans are left with the highest medical costs in the world but with care that often is not based on the best of medical science.

THE TRIUMPH OF COMMERCIAL AUTHORITY

Well-designed, well-conducted, well-analyzed, and well-reported randomized controlled clinical trials are — or should be — the basic unit of scientific evidence that doctors rely on. Over the past four decades, however, increased commercial funding of medical research and the opportunity to turn discoveries funded by the federal government into financial gain for academic institutions and researchers led to exactly the outcome that Harvard's President Bok had warned of. The fundamental goal of academic medical research changed. The pursuit of not-for-profit knowledge in the service of public gain morphed into the pursuit of profitable knowledge in the service of private financial gain.

Despite that, it still seems reasonable to expect that the core precepts of science that protect the integrity of medical knowledge — "reproducibility, rigor, transparency, and independent verification" — would remain in place. But that is not the case. The norms of medical science now require turning a blind eye to the routine violation of the scientific standards that were established more than three and a half centuries ago. In the early 1660s, the Enlightenment-era scientists who established the Royal Society of London adopted a motto that elegantly summarized the key precept underlying their scientific approach: *Nullius in verba*, meaning "Take nobody's word for it." They implored their colleagues to "withstand the domination of authority and to verify all statements by an appeal to facts determined by experiment."

The science behind clinical medicine has regressed from this Enlightenment standard. The sources of knowledge that doctors are taught to trust — peer-reviewed reports of clinical trial results published in medical journals, systematic reviews and meta-analyses, and clinical practice guidelines — routinely violate the basic precept of the Royal Society, rendering the scientific evidence susceptible to commercial bias. Again, this is not to say that *none* of the information found in medical journals is important or that *none* of the new therapies provide heretofore unavailable benefit. Rather, it is a warning to doctors that the primary purpose of the knowledge produced by commercially sponsored clinical research is to serve as a private rather than a public good.

Why doctors are so willing to ignore this warning, leaving them and their patients vulnerable to the manipulations of the drug industry, is the subject of the next chapter.

7

MANUFACTURING BELIEF

Modern economics inherently fails to grapple with deception and trickery. People's naïveté and susceptibility to deception have been swept under the rug.

— NOBEL PRIZE–WINNING ECONOMISTS GEORGE AKERLOF AND ROBERT SHILLER, *Phishing for Phools*

That drug companies bias the journal articles reporting their clinical trial results and the clinical practice guidelines based on those reports is not in question. Why highly trained American doctors fall for this crass commercialism definitely *is* in question.

Part of the answer, discussed in the previous chapter, lies in the false premise that the discipline of evidence-based medicine ensures that doctors will rely on the most valid medical science. The rest of the answer can be found in Big Pharma's marketing toolbox. Therein lie selling tactics rooted in behavioral economics designed to exploit consumers' (including doctors') vulnerability to commercial marketing ploys. Therein also lie the twin tactics of delivering commercially favorable but not necessarily scientifically accurate key messages to health-care providers and creating branding campaigns directed at both consumers and prescribers that are carefully designed to establish an emotional connection to specific products.

Before looking at specific examples of how these tactics have been used, it's important to remember one point discussed in the previous chapter: The FDA's approval process requires evidence that a drug is safe and effective for its approved indication(s) but does *not* require a manufacturer to compare a new drug to the best available therapy. Moreover, the approval process for medical devices such as the implants used in hip-replacement surgery often requires even less evidence. This is due to a loophole in the FDA's process for approving medical devices, the "510(k) pathway." A new device considered "substantially equivalent" to a device already approved can enter the market without any clinical trial evidence, even if that "predicate device" has been taken off the market. In other words, neither drugs nor devices require testing against the best products on the market, and many high-risk medical devices don't require *any clinical testing at all.*

That was the case with Pinnacle metal-on-metal hip implants, manufactured by DePuy Orthopedics, a subsidiary of Johnson & Johnson. (Metal-on-metal hip implants are designed with a metal ball replacing the femoral head that fits into a metal-lined socket placed in the pelvic bone.) By the time the dust settled, more than ten thousand lawsuits had been filed against DePuy, alleging that, contrary to their marketing claims, contact between the two metal surfaces of the joint during routine activity released metal ions into adjacent tissues, causing a destructive reaction. The lawsuits also alleged that DePuy's deceptive marketing had exaggerated the benefits and minimized the risks of its hip implants. Among the few cases against DePuy that actually went to trial, the largest award was made to six plaintiffs from California (grouped into a single trial) who developed complications after receiving DePuy Pinnacle metal-on-metal hip implants.

Two central questions were addressed in this trial: First, were DePuy's metal-on-metal implants more dangerous than other alternatives? (The primary alternative was an older implant that had a metal femoral head inserted into a polyethylene, meaning plastic, socket.) And second, if DePuy's metal-on-metal hip implants were more dangerous, how had orthopedic surgeons been convinced, *despite having not a shred of evidence from randomized controlled clinical trials,* that these implants would be better for their patients?

One of the plaintiffs' orthopedic surgeons — I'll call him Dr. Smith — explained in his testimony why he believed that DePuy's metal-on-metal implant would be best for his "overweight or very active" patients. Dr. Smith had spent hours in his office with DePuy's regional sales manager, who, he recalled, "showed me brochures [illustrating] how the design of the implant allowed fluid film to form between the two metal surfaces [the femoral head and the liner of the socket] and told me the computer design of the implant made all the difference in terms of allowing the fluid fill and getting proper tolerances and sizes and shapes." If Dr. Smith had not been assured that this fluid-film lubrication between the metal surfaces would be maintained, he would have been concerned that over time, direct contact between the metal surfaces would release toxic metal ions into the surrounding tissues. To reassure surgeons like Dr. Smith about the longevity of these devices, DePuy sales reps pointed to the results of the company's own study, the PIN study, which, they claimed, showed that the device had near-perfect performance: 99.9 percent were still in place and functioning five years after hip replacement surgery.

The six-plaintiff case went to trial in 2016. The lawsuit accused DePuy of being negligent in the design of its Pinnacle metal-on-metal hip implants and, further, of knowingly or recklessly making false representations about their advantages and failing to warn patients and doctors of their risks. Before the trial was scheduled to begin, the plaintiffs' attorneys offered to settle the six cases for a total of $1.8 million — $300,000 for each plaintiff. The defendant, Johnson & Johnson, rejected the offer, and the trial went forward.

What the jury heard about the results of the PIN study — the real results — turned out to be quite different from what orthopedic sur-geons had been told in their offices and at professional meetings. DePuy had claimed that its "acetabular cup system" — meaning the socket that is fixed into the pelvis plus the socket liner into which the ball joint is inserted — lasted five years or longer in 99.9 percent of patients. But I had served as an expert in the early phase of this litigation and was part of a team that had been granted access to DePuy's real data, and our analysis showed their claims were not even close to the truth.

Here's what we found — and what the jury heard. Beginning in 2001, the PIN study started enrolling patients who had received DePuy's Pinnacle hip system, which included the metal-on-metal option. The study protocol defined the primary end point as the five-year survival rate of both components of the acetabular cup system (meaning that neither the cup nor the liner was removed for any reason for *at least five years* after hip-replacement surgery). The protocol stated that no interim analysis was planned, so patient data would represent the estimated five-year survival rate for the acetabular components of DePuy's Pinnacle hip implants.

But DePuy flagrantly violated its own study protocol. Rather than waiting the specified five years, DePuy performed an interim analy-sis of the survival rate of the acetabular cup systems in 2006, when the average time from hip replacement was less than two years. DePuy updated its interim analysis of the PIN study data in January 2007, at which time their records showed that, out of the more than one thousand patients in the study, there had been a total of twenty failures — eight for the cup and twelve for the liner. Not only did DePuy deceitfully present that analysis as the final *five-year* results of the study, but it misrepresented its findings, reporting there had been only a single cup failure and no liner failures over five years.

How could DePuy possibly have made that claim of near-perfect performance for its devices? It simply ignored the twelve liner fail-ures and "failed to locate" seven of the eight cup failures. It then re-ported a single cup failure among all 1,183 patients enrolled in the study and based the claimed survival rate on a statistical projection. The truth was that the total number of patients who had actually been in the study five years or more was not 1,183 but a mere 21. Nevertheless, DePuy continued to use the invalid claim of 99.9 percent *five-year* survival rate from the PIN study as the "fundamental selling point for Pinnacle" in its subsequent marketing efforts.

The same month that DePuy submitted its PIN study results to be presented as a poster at the national meeting of the American Academy of Orthopedic Surgeons (AAOS), it also sent the first *four years* of its PIN study data to the French regulatory authority. But instead of the widely touted 0.1 percent five-year failure rate that DePuy claimed in the United States, the failure rate reported to French regulators was 3.4 percent at four years — forty-two times higher.* Almost ten years later, data from actual experience with DePuy implants showed the lie of its claims to orthopedic surgeons like Dr. Smith: The true failure rate of DePuy's metal-on-metal hip implants turned out to be 4.5 times *greater* than that of its older, tried-and-true, metal-on-polyethylene implants.

There was high drama during the second day of the trial when the plaintiffs' attorney Mark Lanier questioned DePuy's CEO, Andrew Ekdahl. DePuy's lawyers objected to one of their client's e-mail chains being entered into evidence, arguing its "prejudicial effect substantially outweighs its probative value." The judge disagreed, so the jury heard the following unorthodox interchange. Paul Berman, DePuy's director of hip marketing, wrote the first e-mail in the chain, dated August 20, 2008. Berman was addressing a significant threat to DePuy's sales of metal-on-metal hips: "We continue to hear questions about hypersensitivity and [metal] ions from surgeons and sales reps in the US." He wrote that he didn't be-

---

* In the results sent to the French regulatory authority, DePuy reported eleven of the sixteen cup-system failures that had occurred in the first four years of the study. Perhaps the data presented to French authorities were more accurate because the legal consequence of misrepresentation of data sent to regulatory authorities can be much greater than misrepresentation of data in marketing claims.

lieve there was evidence of a real problem. Moreover, he added, the situation might even provide "a significant opportunity to further differentiate" DePuy technology.

The opportunity for differentiation that Berman was talking about would come from a modification of DePuy's metal-on-metal implant, called the "aSphere." The proposal was to engineer a non-spherical head for the ball part of its hip implant that would allow DePuy to claim even more complete fluid-film lubrication with even less direct metal-on-metal contact. DePuy could then market this new product as further reducing the risk of shedding potentially toxic metal ions. Berman highlighted the urgency of this project, saying, "We cannot go into the AAOS meeting without approval of the aSphere."

The next e-mail came from Steve Corbett, DePuy's director of manufacturing, who responded, "As far as making these f'n things, let's just make what we always make and just change the label." Within seconds Berman shot back, "Hey, that's not credo behavior," to which Corbett replied sarcastically: "Ooops, you're right. We should definitely shave a rch off of these heads and charge more before we understand what drives free metal ions into the bloodstream."

Before the jury could understand the meaning of Corbett's e-mail, they had to know what *RCH* stood for. After a preemptive apology from both sides, Lanier told the jury, "And when he talks about shaving an rch off, that is a pejorative vulgarity for basically nothing of any count at all. Without doing the entire abbreviation, it's a red C hair."

Lanier then got to the real point: The true vulgarity was not the crude language but the DePuy executives' glib exchange about misleading surgeons into believing their aSphere hip implant was a significant engineering advance without substantial evidence. "Aside from the vulgarity, the concept is unacceptable that you're just going to pretend you've shaved something off of this and sell it for more and claim you've solved a problem? That's vulgar, isn't it?"

Only one of the six plaintiffs in this trial had received an aSphere implant, but all six had been the victims of DePuy's manipulation of the data and callous disregard for the potential of its metal-on-metal hip implants to cause harm. These implants were not only *not* an improvement over the good old-fashioned ones with metal heads and polyethylene liners, which had been available for decades; they were *worse*. They did, however, offer DePuy the opportunity to gain market share by convincing orthopedic surgeons that the metal-on-metal implants would better serve patients who were overweight or who were younger and more athletic.

Only because these cases went to trial do we get to look at the disregard for customers' welfare that was shared so glibly by people in important corporate positions. The trial exposed DePuy's willingness to ignore or reengineer its "scientific evidence" in the interests of selling more product. But it also showed that the corporate greed didn't stop there. To pretend you've shaved something off a product, sell it for more, and claim you've solved a problem is not just vulgar — it's morally corrupt.

Apparently the jury agreed.

Instead of the $1.8 million settlement that Johnson & Johnson had rejected before the trial, the jury awarded the six plaintiffs almost six hundred times more. The $1.04 *billion* award included $30 million for actual damages (such as medical bills, lost income, and physical and emotional pain and suffering) and $1.01 billion for punitive damages (assessed for the defendant's reckless or negligent behavior). The judge later reduced the total award to $543 million — or $90.5 million for each of the six plaintiffs (the details of the final settlement between the plaintiffs and Johnson & Johnson are protected by a nondisclosure agreement).

But a question remains: Why was DePuy's unsubstantiated marketing of its metal-on-metal hips so successful? Couldn't doctors, given their training, skill, and commitment, have seen through DePuy's commercial hype and helped their patients arrive at optimal decisions? After all, that's exactly what is required of physicians in fulfillment of their legal and moral obligation as "learned intermediaries," serving as the interface between medical science and their patients' best interests.

In defense of Dr. Smith and all the other well-meaning orthopedic surgeons like him, medical devices — even high-risk devices like cardiac pacemakers and joint implants — are often approved by the FDA through the 510(k) pathway described above, which does not require preapproval clinical studies demonstrating safety and efficacy. As might be expected, this lax regulatory oversight is associated with an increased risk of unforeseen safety issues: Medical devices approved without clinical trial data are eleven times more likely to be recalled by the FDA than those approved through the conventional process. DePuy's Pinnacle metal-on-metal hip implants were approved by the FDA through this 510(k) process. And the first so-called clinical evidence available to doctors about the survival rate of DePuy's metal-on-metal hip implants was the grossly misrepresented implant-survival data from the PIN study.

Turning back to prescription drugs, the U.S. Government Accountability Office reported that between 2005 and 2016, "Novel drugs — innovative products that serve previously unmet medical need or help advance patient care — accounted for about 13 percent of all approvals each year." With only about one out of eight new drugs truly advancing patient care, American doctors, insurers, health-care economists, government officials, and the public are left to fend for themselves in the impossible task of determining the role of new products in optimal care. I say "impossible task" because our system of privatized clinical research and manufacturers' control over the communication of most of the results creates an impregnable wall around the information. Economists call this "asymmetry of information"; the rest of us call it hidden information or corporate secrets.*

---

* Although there has been some movement toward making more data available, there has been none toward making data available prior to the publication of study results in medical journals, the point at which doctors formulate opinions about a new drug or other product.

As shown in the previous chapter, corporate sponsors of clinical trials typically design and conduct the studies, own the data, perform the analyses, participate in the development of articles for medical journals, and direct the marketing of their products. This places those corporations in the commercially advantageous position of knowing a whole lot more about the true value of their drugs and devices than do the doctors who prescribe them. The resulting asymmetry of information about prescription drugs allows manufacturers not just to spin clinical study results favorably but to get away with selectively withholding or misrepresenting their data. And this in turn allows — in effect, mandates — that manufacturers fully cash in on this unprecedented marketing opportunity.

The most important lesson I learned during my years serving as an expert in litigation was not about the particularities of any specific drug company's wrongdoing buried in corporate computers, although these were numerous and often of major consequence. Rather, it was how predictably doctors were influenced by the drug companies' biased presentation of their "scientific evidence" and their marketing campaigns — just as I had been, as a hardworking family practice doctor, especially earlier in my career.

Hence my purpose in writing this book: to show how, by the skillful use of the tactics already listed and to be discussed in the next three sections of this chapter, pharmaceutical firms influence the behavior of health-care professionals. No longer the learned intermediaries our patients expect us to be, we have unwittingly become "unlearned" intermediaries, serving the drug companies' financial interests rather than our patients' medical needs.

BEHAVIORAL ECONOMICS AND DOCTORS

To understand why doctors are such easy marks, we need to take a closer look at their vulnerabilities as consumers of commercially produced scientific information. This requires going beyond the paradigm of evidence-based medicine to the psychology that lies behind doctors' processing of new medical information. That psychology leaves them just as susceptible as ordinary consumers are to manipulation by skillful marketing.

In 2013, two Nobel Prize–winning economists, George Akerlof and Robert Shiller, were writing *Phishing for Phools: The Economics of Manipulation and Deception.* Their book debunks the myth of the utopian market being magically overseen by an "invisible hand." In it, the authors state that Adam Smith's unregulated market is not in fact a mechanism for optimally distributing goods and services to the maximum benefit of society; rather, it is a way to exploit human weaknesses (and thus engage in "phishing"). And this is possible because consumers often confuse rational needs and preferences with unconscious desires (like people who choose to purchase a particular car without being aware that the appeal derives from its aura of power or wealth), which renders them vulnerable to exploitation by skillful marketers (and thus destined to act as "phools").

This framework sheds light on the special case of how doctors acquire their knowledge about optimal patient care. A doctor's fundamental act as a consumer is not the decision to write a specific prescription for a specific patient. More precisely, it's the decision to accept that industry's "scientific evidence" about a drug or device fulfills his or her criteria for personal adoption of a professional belief. Analogous to the moment when conventional consumers decide to make a purchase, doctors decide whether new information merits being accepted as knowledge and is therefore worthy of being integrated into their repertoire of optimal patient care. The more effective the drug company's marketing, the more difficult will be the doctor's decision *not* to accept the supposedly authoritative scientific evidence. And this goes even for doctors' encounters with drug reps, who usually have minimal scientific background but are armed with authoritative-appearing marketing materials (including reprints from medical journals) and trained to present information in the way that doctors have been trained to receive it. Thus, the drug companies aim about two-thirds of their marketing budget at doctors and other prescribers (most of the rest is advertising aimed at consumers).

Professor Akerlof had read *Overdo$ed America* and found my description of the deception perpetrated by the drug companies "an extremely good example of the phenomenon" of drug companies doing "phishing" and doctors tending to act as "phools." He e-mailed me with an inquiry: "I am trying to parse out whether the activities of Bad Pharma are a serious pervasive problem, or instead are the sins (which no doubt should be monitored and corrected) that would be expected to occur in any activity that involves so many people who naturally need to make profits to support their activities." He was particularly interested in whether critiques of the pharmaceutical industry's excesses, offered by myself and others, had ever been meaningfully rebutted, noting that economists who are more supportive of the industry had not "done the careful parsing of the medical literature that you have done." I wrote back that in the nine years since my book's publication, I had not been informed of any significant errors.

Akerlof and Shiller's book, published in 2015, describes drug company marketing as "gaming the doctors," the purpose of which is to "create the story of the new wonder drug." The drug companies approach this story formation from two directions. The first involves understanding how certain behaviors (gleaned from the tradition of behavioral economics) increase people's susceptibility to marketing manipulation — that is, their "phishability." These behaviors (adapted from social psychologist and marketer Robert Cialdini) include

- having a sense of reciprocal obligation after receiving gifts and favors
- wanting to be polite, especially to those who treat us with kindness or respect

- tending to avoid disobeying authority
- wanting our behavior to be consistent with the expectations of others
- wanting our decisions to be internally consistent
- acting to avoid economic, social, or interpersonal losses

From the perspective of evolutionary psychology, these tendencies facilitate humans' living peacefully in stable social groups. But on the flip side, they render people vulnerable to manipulation by leadership (and now by marketing) that prioritizes its own selfish interests over the group's welfare.

This list was not made with doctors in mind, though it certainly could have been. Drug reps exploit doctors' sense of reciprocal obligation by leaving small gifts and buying meals, which are accepted by 70 percent of practicing physicians each year. Even a single meal worth no more than twenty dollars provided by a drug company significantly increases doctors' prescribing of brand-name drugs that are no more effective than drugs of the same class available as low-cost generics. A review article found *all the studies* that explored the effect of industry gifts and payments showed a positive effect on physicians' prescribing. Drug reps are trained to engage doctors in encounters by storing personal information about their interests, children, and marital status and by using scientific language to sound knowledgeable. Once doctors are engaged, the reps can then deliver their marketing department's well-honed messages designed to increase use of the products they represent. From personal experience, I know that avoiding these commercial propaganda sessions requires a readiness to be direct and sometimes uncomfortably impolite.

More broadly, by their selection and training, doctors are reluctant to contradict authority and don't want to incur the wrath of peers, oversight organizations, or lawyers poised to sue if quality-of-care standards are not followed and there is a bad outcome. Doctors' acceptance of commercially influenced evidence published in journals and the recommendations in clinical practice guidelines (or the marketing materials based on them) is not just the path of least resistance; it's the only path that will allow them to get through their professional day.

But it's Akerlof and Shiller's second approach to story formation that goes to the core of explaining why people, including doctors, are so vulnerable to making decisions that contradict their avowed self-interest. The key element is marketing designed to blend the self-deceptive stories people tell themselves with skillfully fabricated stories about commercial products, coaxing consumers (including doctors as consumers of information about the benefit of new drugs) toward fulfillment of marketers' goals rather than their own.

Akerlof and Shiller define a subcategory of "information phools," meaning those who "act on information that is intentionally crafted to mislead them." Bull's-eye. The stories that make doctors vulnerable to being information phools are not hard to imagine. Doctors could not practice without believing they are using the best of medical science to serve their patients, and, like everyone else, they want to be respected by their peers and authority figures. To these ends, doctors strive to practice evidence-based medicine and are loath to jeopardize the professional and social standing they have worked so hard to achieve. This leaves most doctors with a strong antipathy toward rocking the commercially powered evidentiary boat.

Because of this epistemological blind spot, adherents of the paradigm of evidence-based medicine remain oblivious to the routine violation of the most fundamental requirements of valid science: transparency of data and confirmation by independent analysis. Doctors naively assume that the peer-reviewed scientific evidence on which they rely satisfies these basic requirements. But as we have seen, it does not, rendering doctors exquisitely vulnerable to drug company phishing. And, as shown in the next section, also highly vulnerable to the key messages that marketing departments so skillfully disseminate.

KEY MESSAGES DESIGNED TO MISLEAD

As an expert witness in litigation, I have had access to manufacturers' scientific data as well as their business and marketing plans. These are the pieces of the puzzle that, when put together, reveal the tactics drug companies use to persuade doctors to prescribe their expensive new drugs even when they offer little or no added value (and sometimes harm) compared to less expensive alternatives.

Doctors receive information about new products primarily through medical journals and, later, in clinical practice guidelines; continuing medical education lectures and marketing material from drug and device companies also play their part. The manufacturers use each of these sources like a Trojan horse to deliver "key messages" that have been skillfully honed to convince doctors that prescribing the latest drug is in their patients' best interests. The primary purpose of these key messages is *not* to communicate accurate and complete study results. Instead, they are road-tested marketing messages, Pharma-designed for effective phishing of doctors to maximize sales. Less generous terms for many of these key messages would be *exaggerations, expertly developed commercial propaganda,* and, sometimes, simply *lies.*

A confidential corporate document produced in the Neurontin litigation discussed in chapter 2 provided a rare window into how key messages are used to optimize the marketing impact of the so-called scientific evidence delivered to doctors. The slide shown on the next page, titled "Aligning Marketing and Scientific Key Messages," was prepared for Pfizer by Medical Action Communications (MAC) as discussed in chapter 2. MAC's slide instructs Pfizer on how to use marketing-generated key messages to shape, rather than reflect, the scientific evidence delivered to physicians. The top half of the slide shows how, without MAC's assistance, Pfizer's marketing efforts would produce the undesired outcome of "Inconsistent Key Messages" between marketing material ("Proceedings or

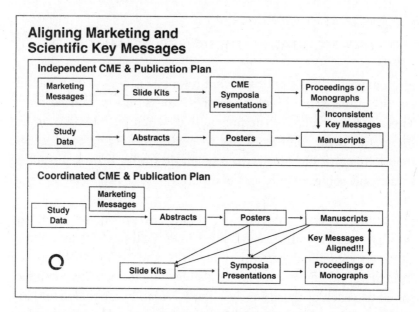

Aligning marketing and scientific key messages *Slide from September 2002 presentation. Trial Ex. 259 in Kaiser v. Pfizer.*

Monographs") and study data published in medical journal articles ("Manuscripts").

In the bottom half of the slide, MAC shows Pfizer how to avoid this problem. Prior to study data being integrated into the manuscripts published in peer-reviewed medical journals, the data must first pass through a filter to make certain the reported clinical trial results support the predetermined "marketing messages." This ensures that the results published in medical journals deliver the same takeaway points as the corporate marketing materials. The result: "Key Messages Aligned!!!" Complete and accurate communication of scientific evidence from clinical trials be damned.

BRANDING .

Key messages are integrated into branding strategies that are then used by drugmakers to communicate the purported therapeutic and emotional benefits of their products. Health-care professionals delude themselves that branding efforts targeted at consumers rely primarily on emotional connection through storytelling, whereas branding directed at them is based on rigorous scientific evidence. As shown in the next two sections, the branding strategies for the enormously successful marketing of Nexium and OxyContin were based on stories told to doctors that effectively captured their beliefs even though the claimed superiority of the products was *directly contrary* to the manufacturers' own scientific evidence.

Nexium — the Purple Pill

In 2000 the stomach-acid-blocking medication Prilosec was more than just AstraZeneca's best-selling drug. With over $6 billion in annual sales, Prilosec was the best-selling drug in the world. Needless to say, the impending expiration of the patent on Prilosec the following year posed a major threat to AstraZeneca (AZ). That's why, back in 1995, it had commenced the "Shark Fin project," so named because of the upside-down V that the graph of corporate sales would look like if AZ failed to find an adequate replacement for Prilosec. The company chose Nexium, even though its researchers knew early on that Nexium would not provide an advantage over Prilosec for its most common use, treatment of heartburn, and considered it to be "among the poorest" of the possible choices. But Nexium provided one important advantage: It gave AZ the opportunity to create the *appearance* of superiority.

When the FDA medical officer reviewed the new drug application for Nexium in 2000, he concluded that AZ's claim that it provided "a significant clinical advance over [Prilosec] in the first-line treatment of patients with acid-related disorders **is not supported by data**" (boldfaced in the FDA report). And a little over a year after Nexium entered the market, Thomas Scully, then head of the federal Centers for Medicare and Medicaid Services, told doctors at an AMA conference, "You should be embarrassed if you prescribe

Nexium" because it was so much more expensive but no more effective than generic alternatives.

He was right. Nexium provided no advantage over omeprazole (the generic name for Prilosec), yet it cost about eight times more. Nevertheless, AZ's branding strategy was so successful, it created the perception that Nexium was an innovative, superior therapy, using terms that appealed to both prescribers (who wanted to provide the best care for their patients) and patients (who wanted to obtain what TV ads implied was the best available therapy). In fact, AZ's marketing was so effective that Nexium's launch became the most commercially successful drug launch ever. And by 2017 AZ had persuaded so many doctors to prescribe Nexium that it had become the third-highest revenue-generating drug of all time (bested only by Lipitor and Humira), racking up $72 billion in U.S. sales.

This is how the ruse worked. Organic molecules — like omeprazole — usually come in two mirror-image versions (called enantiomers), like hands placed palm to palm. Prilosec is an equal mix of both enantiomers of the omeprazole molecule, whereas Nexium is composed of only one of the two enantiomers. Both enantiomers in Prilosec and the single enantiomer in Nexium are transformed into the same molecule when they reach their target, the acid-producing cells in the stomach. The only difference between Nexium and Prilosec is that the single enantiomer in Nexium is metabolized more slowly than the two enantiomers in Prilosec, which means patients taking Nexium have higher blood levels of active drug than patients taking an equal-milligram dose of Prilosec. AZ's multibillion-dollar sleight of hand was that, rather than adjusting the milligram dose of Nexium *downward* to account for its slower rate of metabolism, the company *increased* the milligrams in the typical dose (from 20 milligrams to 40), resulting in three times the blood level reached by 20 milligrams of Prilosec — and then claimed that Nexium was more effective.*

---

* The "tell" here is that AZ priced the 40-milligram Nexium *lower* than the 20-milligram Nexium. In the rare cases where Nexium was more effective

You may be wondering why AZ deserved a new patent for Nexium, which gave it the opportunity to create a whole new brand with another round of monopoly sales, when the drug was simply half of the active ingredient already inside the Prilosec capsule and its effectiveness, based on equivalent blood levels, was identical. I served as an expert in litigation about this very issue. Had the plaintiffs — unions, insurers, and individuals who paid for Nexium — won this case, AZ would have had to reimburse them for all or part of the approximately eight times higher price they paid for the now generic Prilosec. As it happened, the case was settled, so the confidential documents remain sealed. But several FDA documents that I analyzed for litigation tell most of the story, and this I can share.

The first step in AZ's marketing campaign was to bias the scientific evidence that physicians received from peer-reviewed medical journals. This was accomplished by the selective and timely publication in medical journals of only two out of the seven completed clinical trials included in its 1999 new drug application for Nexium that compared Nexium to Prilosec. Both of these published studies enrolled patients with erosive esophagitis (erosion of the esophagus diagnosed by endoscopy and usually caused by acid reflux) and showed a significant advantage of a daily dose of 40 milligrams of Nexium compared to 20 milligrams of Prilosec — in effect, three times the dose of the latter. But neither study, as reported by the FDA medical officer, provided evidence of the superiority of 20 milligrams of Nexium over 20 milligrams of Prilosec. These two so-called positive studies were published in 2000 and 2001, coinciding with Nexium's entrance into the market, which was when doctors would be forming their opinions about the new drug. The other five studies, all of which failed to show that Nexium was superior, were not published until 2004 and 2006. By then doctors, patients, and purchasers had long ago formulated their beliefs based on the initial, strategically curated "scientific evidence" that had been cho-

---

than an equivalent milligram dose of Prilosec, the dose of the latter could simply have been increased to obtain the same result.

sen to convince them of the superiority of Nexium compared to Prilosec.

The second step in AZ's marketing campaign was to leverage the
biased scientific evidence that was available to persuade doctors to
stop prescribing Prilosec and start prescribing Nexium. The company had a big head start with its six thousand drug reps who had
ongoing relationships with doctors who had been prescribing Prilosec. Those perfectly positioned and retrained drug reps delivered
more sales pitches for Nexium than doctors received for any other
stomach-acid-reducing drug. The *Wall Street Journal* described
how an AZ drug rep made the claim that Nexium was better than
Prilosec by telling the doctor, "The proof's in the healing rates," and
then citing the results of the studies that compared the higher dose
of Nexium to the typical dose of Prilosec.

And the third step in AZ's marketing campaign was to conduct
the most intense direct-to-consumer (DTC) advertising campaign
in history. Between 2002 and 2006, AZ spent a record $1.08 bil-
lion on advertising Nexium. The ads were crafted to create the impression of the superiority of Nexium without actually making that
claim.

There is, however, one area in which the typical dose of Nexium *is* superior to Prilosec: in creating the need for itself. The rate
of production of stomach acid is controlled by a hormone called
gastrin, blood levels of which normally rise after eating to stimulate secretion of the stomach acid needed for proper digestion.
When acid formation is suppressed by Nexium or Prilosec, blood
levels of gastrin increase in proportion to the extent of acid suppression as the body tries to stimulate the cells in the stomach to
produce more acid; it's like a thermostat sending a signal to the
furnace to switch on when the room temperature has fallen below the set point. After two or three months of Nexium therapy,
gastrin levels reach a maximum, and it takes up to four weeks following discontinuation of the drug for gastrin levels to return to
normal. This leaves many people who stop taking Nexium with
symptomatic gastrin-stimulated overproduction of stomach acid.

This so-called acid rebound can create intolerable symptoms that require — you guessed it — restarting Nexium. Almost one out of three healthy volunteers given 40 milligrams of Nexium once daily for eight weeks developed acid-rebound symptoms after stopping it.

The moral of the Nexium story is *not* that AstraZeneca's misleading marketing made it an evil outlier among drug companies. Quite the contrary — the point is that AZ used journeyman tactics to create brand value by exploiting doctors' informational vulnerabilities (that is, their paradigm of legitimate sources of information) and the public's susceptibility to DTC marketing. AZ was certainly not alone in using this strategy; it was just more successful. And such marketing is by no means limited to the drug companies. In the case of metal-on-metal hip implants, we saw that DePuy created brand value in an unproven medical device that turned out to be far less effective than its predecessors.

Finally, if we take a step back from the specifics of AZ's marketing, we see that the value of AZ's patents on Nexium — as with most new drugs that are no more effective than drugs already on the market — did not reside in the chemical itself. Rather, it was as if AZ had been granted a patent on the contents of an empty box labeled NEXIUM that it could then fill with whatever marketing claims it thought it could get away with to convince doctors of the nonexistent advantages of Nexium over Prilosec. As psychiatrist and psychopharmacologist David Healy explains, prescription medicines are "chemicals plus information," and much of what is now taken for valid information about brand-name drugs (from even the most trusted sources) is "fake information."

Thus, AZ's patents on Nexium allowed the company to invest heavily in filling the box with whatever marketing malarkey they wished, as long as those claims stayed within the letter of the law. If AZ had not owned those patents, other companies could have marketed the single enantiomer in Nexium and negated the value of AZ's investment in its deceptive, if not outright dishonest, branding campaign.

OxyContin — Sold as Long-Acting Pain Relief
with Little Risk of Addiction

Nexium had the built-in marketing advantage of creating the need for itself, but the grand prize for misleading doctors with the most malevolent prescription-drug-branding campaign for a drug that created the need (and desire) for itself goes to Purdue Pharma for its marketing of the opioid OxyContin as low-risk, long-acting, less addictive, and less likely to be abused than other narcotics. This saga started in 1984 when Purdue first marketed MS Contin, a long-acting formulation of age-old morphine sulfate in a wax matrix taken every eight to twelve hours to treat cancer pain. In 1990, Purdue's success with MS Contin brought the company to the edge of the same precipice that AZ would face a few years later: preparing for the expiration of the patent on its primary revenue generator.

Purdue more than met the challenge. In 1996 — the year its patent on MS Contin expired — Purdue transferred its marketing efforts from morphine to another opiate: semi-synthetic oxycodone, which had been developed in Germany back in 1916. Purdue's marketing of its new long-acting formulation of oxycodone, brand name OxyContin, claimed several advantages (or created a false impression of these advantages, which it left uncorrected) over other oral narcotics:

- that oxycodone was a *less* powerful narcotic than morphine and could therefore be safely used to treat non-cancer-related chronic pain;
- that using a wax matrix to control the release of oxycodone from OxyContin tablets slowed systemic absorption of the active drug, which greatly decreased the risk of abuse, addiction, and withdrawal symptoms; and
- that OxyContin would provide round-the-clock pain relief when given just twice daily.

Purdue engaged in an unusually aggressive marketing program. It repurposed its cadre of drug reps — whose job had been to increase sales of MS Contin — to "educate" doctors about OxyContin being less addictive and less likely to be abused than other prescription opioids. Purdue paid all the expenses for training more than five thousand doctors, pharmacists, and nurses to give lectures (at sunny resorts) promoting OxyContin. And between 1996 and 2002, Purdue funded over *twenty thousand* educational meetings on pain control for health-care professionals.

In addition to this explicit marketing, Purdue funded nonprofit organizations like the American Pain Society, the American Academy of Pain Management, and the American Geriatric Society to recommend treatment of non-cancer chronic pain with opioids. In 1995 the American Pain Society issued quality-improvement guidelines for the treatment of pain that recommended *level of pain* be evaluated as a vital sign in relevant patients. In 2001 the Joint Commission on Accreditation of Healthcare Organizations (JCAHO) adopted the recording of pain as a fifth vital sign as a standard requirement for hospital accreditation. In 2001 and 2002, Purdue sponsored a series of nine educational programs designed to teach hospital-based doctors and staff how to implement the new standards for pain control. Purdue's agreement with JCAHO made it the only drug company permitted to distribute specified educational materials within hospitals.

Dr. Abigail Zuger, writing in 2018 in the *New York Times Book Review*, eloquently explained why so many physicians prescribed OxyContin: "Granted, a few of us were criminals, methodically defrauding Medicaid in pill-dispensing 'mills.' But mostly we were just well-intentioned schlubs with prescription pads, dutifully following then-current practice guidelines." And why shouldn't these "well-intentioned schlubs" have dutifully followed those guidelines? They perceived a professional duty to do as the guidelines recommended, even though the highly respected professional organizations that issued those guidelines had received generous —

but not always forthrightly declared — funding from opioid manufacturers. And Purdue led the list.

By 2001, OxyContin had become the most frequently prescribed brand-name narcotic in the United States, with annual sales reaching $1 billion. This was a remarkable marketing feat, given that the three claims Purdue had made about the advantages of OxyContin were not true. In the first place, oxycodone is not weaker than morphine; it is actually *twice as strong.* Second, Purdue had not done any studies to support its claim that the controlled-release formulation of OxyContin using a wax matrix made it less likely than other narcotics to lead to addiction. Notwithstanding this lack of evidence, the FDA allowed the following unsubstantiated claim to be included in the label for OxyContin: "Delayed absorption, as provided by OxyContin tablets, is believed to reduce the abuse liability of a drug." The U.S. Drug Enforcement Agency, however, described the effect of OxyContin, when the wax matrix was crushed and the contents snorted or dissolved in water and injected, as similar to that of heroin. Indeed, a study done by Purdue showed if OxyContin tablets were crushed and stirred into water, two-thirds of the active narcotic could be drawn up into a syringe for intravenous injection.

But probably the most insidious of Purdue's claims was the promise of round-the-clock relief if the drug was taken every twelve hours. Within a year of OxyContin's introduction to the market, Purdue was receiving feedback from physicians that pain relief was *not* lasting twelve hours, leading some doctors to prescribe it every eight hours. A Purdue regional sales manager responded to these complaints with a memo instructing drug reps to tell doctors "that 100% of the patients in the studies had pain relief on [a twelve-hour] dosing regimen." But Purdue knew this was not true. Its very first study of OxyContin, begun in 1989, had enrolled ninety women who had undergone abdominal and gynecological surgery. More than a third of the women experienced the return of post-op pain within eight hours of receiving OxyContin, and about half needed additional medication within twelve hours.

Why didn't Purdue simply acknowledge the need for more frequent dosing in many patients (and sell more pills in the process)? The company had several reasons to be untruthful about the duration of pain relief: insurers balked at filling prescriptions for OxyContin for more frequent than twice-daily dosing, arguing that far less expensive shorter-acting medications could be used instead; doctors found the simplicity of twice-a-day dosing an advantage; and unlike many other drugs, OxyContin pills were priced in proportion to the dose of the active ingredient, oxycodone, incentivizing drug reps to recommend that doctors increase the *dose* rather than the *frequency* of OxyContin.

This strategy may have been designed to maximize OxyContin sales, but, according to Theodore J. Cicero, neuropharmacologist at the Washington University School of Medicine in St. Louis, it was a "perfect recipe for addiction." Not only were most patients experiencing pain toward the end of the twelve-hour dosing period, but many were also experiencing narcotic-withdrawal symptoms. Increasing the dose rather than the frequency brought in more money, but it also created more addiction and raised the risk of fatal overdose. Purdue, by rigidly maintaining the twelve-hour dosing schedule and increasing the total daily dose of OxyContin, ensured that the drug did exactly the opposite of what it was supposed to do: control pain without the risk of addiction.

In 2007 Purdue Pharma pleaded guilty to a felony charge of illegally misbranding OxyContin "with the intent to defraud or mislead physicians and consumers." Purdue agreed to pay $600 million in criminal and civil penalties and "acknowledged that it illegally marketed and promoted OxyContin by falsely claiming that OxyContin was less addictive, less subject to abuse and diversion, and less likely to cause [w]ithdrawal than other pain medications — all in an effort to maximize its profits."

Paradoxically, OxyContin sales shot *up* after Purdue pleaded guilty to making these felonious marketing claims. Perhaps this was in part due to the efforts of McKinsey, the consulting firm that Purdue hired to help increase sales of OxyContin. A 2008 e-mail written by a McKinsey consultant recounts that a letter from the Purdue board "'blessed' him to do whatever he thinks is necessary to 'save the business.'" And the business was indeed saved! U.S. sales grew to $3 billion in 2010. Meanwhile, the number of U.S. deaths due to overdoses of prescription opioids increased roughly in proportion to growing OxyContin sales, from 3,400 in 1999 to about 14,000 in 2010.

In 2010 the FDA approved a reformulation of OxyContin that prevented the drug from being snorted or injected (Purdue added a hard coating on the outside of the tablets, which made crushing difficult and caused the contents to form a gel when water was added). Perhaps Purdue was motivated to reformulate the drug at that time in part because the reformulation allowed it to extend its patent, scheduled to expire in 2013, all the way to 2030. Deaths from prescription-opioid overdose leveled off at that point but did not decline.

Then the nature of our opioid problem took an even more deadly turn. The next generation of opioids were not prescription drugs, and the dose was not controlled. Inexpensive illegal fentanyl — a synthetic opioid described by the U.S. International Trade Commission as "50 times more powerful than heroin and 100 times stronger than morphine" — started to enter the United States from China. It is used as a cheap additive to heroin, cocaine, and methamphetamine and made into pills sold on the street. Overdose deaths involving illegal fentanyl have risen sharply, from about 3,000 in 2013 to more than 36,000 in 2019.

The economic and social conditions that fostered the increasing "deaths of despair" described in chapter 5 combined with Purdue's success at enlisting physicians to normalize the widespread use of prescription opioids and the introduction of inexpensive fentanyl into street drugs to create a full-blown disaster. According to the CDC, prescription and illegal opioids claimed more than 450,000 American lives between 1999 and 2018.

The President's Commission on Combating Drug Addiction and the Opioid Crisis was formed in 2017. Professor Bertha Madras of

Harvard Medical School, the commission report's principal author, said the drug companies invested "massive amounts of money" to influence doctors' prescribing of narcotics: "Using that money they literally bought off, and I don't use that phrase lightly, they bought off the Joint Commission (which accredits hospitals and sets medical policies), they bought off the Federation of State Medical Boards, they bought off several American pain associations." The commission concluded that Purdue Pharma's marketing campaign for OxyContin resulted in a tenfold increase in prescriptions for opioids and set the stage for the addiction and overdose problems that followed.

Professor Madras further mused on the influence that was bought by the drug industry: "I think that the vast amounts of money were used very effectively. It was a national campaign. Have we learned the lessons? Absolutely not."

As untethered from moral decency and scientific responsibility as Purdue's creation of excess brand value for OxyContin was, the company's efforts would have been for naught had doctors not prescribed the drug. As Dr. Zuger observed, the vast majority of doctors who wrote those prescriptions "were just well-intentioned schlubs" who believed they were serving their patients' best interests.

There is plenty of blame to go around in this tragedy, but the primary focus here is on the power of the purposely misleading information doctors received from the channels they had been taught to trust. The drug companies are usually the source of this misinformation, but they would be unable to create excess brand value on their own. They need the assistance of a network of individuals and organizations to produce and deliver the information that fulfills doctors' learned criteria of belief while also fulfilling the drug company's goal of increasing sales. How the pharmaceutical industry incentivizes that network to deliver inaccurate medical information to physicians is described in the next chapter.

8

MARKET FAILURE IN MEDICAL KNOWLEDGE

Information issues are key to many different types of market failure ... And when information is lacking, or hidden, the standard theories of economics ... often don't apply.

— JOHN CASSIDY,
How Markets Fail: The Logic of Economic Calamities,
discussing a key insight of Nobel laureate Joseph Stiglitz

The case against [biomedical] science is straightforward: much of the scientific literature, perhaps half, may simply be untrue. ... Science has taken a turn toward darkness.

— RICHARD HORTON, editor in chief of *The Lancet*

I t was a picture-perfect July day at a posh Cape Cod resort. The summer breeze coming off the water was warm and gentle, the sky a gorgeous blue. I was there to address the 2005 annual meeting of the World Pension Forum. The leaders of the nation's largest public-employee pension funds were in attendance. These fund managers were responsible for investing the retirement savings that millions of public employees were relying on for their financial security and future health benefits.

My talk focused on the extent to which pharmaceutical companies were using aggressive marketing (often misrepresenting the underlying science) to create blockbuster drugs. As an example, I described how a dangerous combination of purposely manipulative marketing and unintentionally misleading epidemiological studies had persuaded doctors to prescribe hormone replacement therapy (HRT) for twenty million healthy American women entering menopause. Large observational studies had shown that women taking HRT were significantly healthier than those who were not taking hormones, with a reduced risk of heart disease likely to increase life expectancy. Industry-funded experts told these women and their doctors that HRT would protect them during the "estrogen deficiency" phase of their lives: "Breasts and genital organs will not shrivel. Such women will be much more pleasant to live with and will not become dull and unattractive."

But when finally studied in a large, publicly funded randomized controlled trial (the Women's Health Initiative study, funded by the National Heart, Lung, and Blood Institute), HRT, it turned out, significantly *increased* women's risk of heart disease, stroke, and breast cancer. I explained that the erroneous conclusion that HRT led to better health was due to a phenomenon known as the healthy-user bias. Whereas the studies appeared to show that better health in women undergoing HRT was due to their taking hormones, what really was happening was just the reverse. The women treated with HRT were more likely to be healthier because they were better educated, wealthier, and receiving more preventive medical care. The truth: Taking HRT was an indication of women's *predisposition* to — and not the *cause* of — their better health.

I then shared how the misreporting of clinical trial results involving Vioxx (Merck) and Celebrex (Pfizer), published in the most respected medical journals, steered doctors away from FDA reviews that showed the lie of the manufacturers' safety claims about both drugs. And I discussed how statin guidelines recommending that doctors prescribe cholesterol-lowering drugs for healthy women and men over sixty-five were scientifically unsubstantiated, perhaps in part because all but one of the nine guideline authors had financial ties to statin makers (which they failed to disclose).

I concluded my talk by observing that the public employees whose welfare my audience was charged with overseeing were not being best served by the current system of drug development and marketing. I noted that doctors were being misled by the drug companies' exaggerated claims of efficacy and minimized reporting of risks of their products, and I ended by explaining that this was playing a significant role in the comparative decline of Americans' health relative to that of other wealthy countries, even while we spend so much more on health care.

I anticipated an enthusiastic response. What I got was more of an awkward thud.*

That evening the conference reconvened for socializing at the quaint and elegant Cape Cod Museum of Art. I made small talk with a few managers of the larger funds, and after a couple of rounds of drinks, I brought up the subject that was on my mind: "Why such an unenthusiastic response to my talk?" The answer was simple. One of the fund managers explained, "The pension funds represented in that room own one and a half percent of both Merck and Pfizer, and we are not so happy to hear what you have to say about the behavior of the drug companies."

I had made the fund managers uncomfortable by placing them inescapably in the middle of the basic dilemma of American health care: Should its primary purpose be providing the most effective and efficient health care? Or should it be maximizing corporate profits, thereby obligating Big Pharma to control (in the service of their own interests, of course) what American doctors, policy makers, and the public believed to be the best scientific evidence?

---

* I received a much more enthusiastic response after delivering a similar talk to the Direct to Consumer Advertisers' convention, although admittedly from the podium it looked like most of the people showing their appreciation were clapping under their tables.

My talk highlighted just how irreconcilable this dichotomy had become. Corporate management and investors (including managers of public-employee pension funds) were acting rationally to fulfill their professional responsibilities, but their success was built on a system that had evolved to deliver Pharma-biased information through doctors' and the public's most trusted sources of information. Big Pharma and its investors were doing just fine financially, but society as a whole was not receiving commensurate value. Economists call this market failure. With regard to prescription drugs, this failure is occurring at several levels.

MARKET FAILURE #1: MONOPOLY PRICING

Four out of five Americans think the price of drugs is too high, with good reason. Between 2008 and 2015, the price of frequently used brand-name drugs rose almost fourteen times faster (164 percent) than the Consumer Price Index. Americans now spend almost twice as much per person annually on prescription medications as the citizens of ten other wealthy countries, $1,443 versus $749, respectively.

As shown in chapter 6, the reason for these high prices is no mystery. We are the only wealthy country that does not have a national mechanism for regulating the price of prescription drugs. Rather, we grant manufacturers market exclusivity* — creating a monopoly on every new drug approved by the FDA — and then allow the drugmakers to invest in creating brand value and to charge whatever price they wish.

An article in the *Journal of the American Medical Association* concluded that each year Americans pay an excess $170 billion for

---

* Market exclusivity is protected by the FDA for five to seven years after approval of conventional drugs and twelve years for biologic drugs. It is also protected by twenty-year patents issued when a drug is first registered, although gaining FDA approval usually consumes six to eight of those years.

their prescription drugs due to "medication pricing failure" — that is, pricing not justified by the value provided. This amounts to an extra $550 transferred annually from every American man, woman, and child into Big Pharma's coffers. And no matter what the manufacturers say, these unregulated prices bear little to no relationship to the actual cost of drug development.

Our uniquely market-based approach to the pricing and distribution of prescription drugs is uniquely inefficient. For starters, the "market" for prescription drugs bears little resemblance to the classical market depicted by Adam Smith back in 1776. In the market Smith described, butchers, brewers, and bakers offered their products to compete with similar products from other merchants. Consumers' direct experience allowed them to make wise purchases while merchants made a fair living. Market regulation was built in: If Baker A's bread was priced too high or was of inferior quality, consumers could buy bread from Baker B.

The market for prescription drugs bypasses these self-regulating relationships in three fundamental ways: Consumers cannot choose which prescription drugs to purchase (health-care professionals must prescribe them); doctors recommending drugs to their patients are insensitive to price (it's not their money); and the discipline of price is largely eliminated since most consumers pay only a small fraction of the cost (because of insurance).

But the most important difference is more subtle. In the market Smith described, consumers directly evaluate the quality and compare the price of material goods being offered, but the primary offering in the market for prescription drugs is not a material product. Rather, pharmaceutical manufacturers market *beliefs* about their drugs to the medical community (and to the public through advertising and public relations efforts), beliefs designed to maximize their sales. Since there is inadequate oversight of the accuracy and completeness of the information behind those sales pitches and no ceiling on what can be charged for the drugs being touted, such aggressive marketing has become not just the norm but a requirement for the manufacturers. In essence, then, Baker A's bread often gets prescribed by doctors (based on their unsubstantiated belief in its superiority) and purchased for a premium price by consumers (heavily subsidized by insurance) even though Baker B's bread would provide more value. Meanwhile, the benefits of other approaches, such as healthy lifestyle changes, are completely ignored in the context of this economically driven competition.

The bottom line is that American consumers are buying new, expensive drugs that are often no more effective than older cheaper drugs and on occasion may actually cause harm. For those who have good insurance, the cost of these drugs is spread across large groups, so the high price exerts little consumer discipline. For those whose coverage is not good or is nonexistent, purchasing these drugs, helpful or not, may bring financial hardship. However, at no time is the ultimate consumer in the drug market (meaning the drug user) ever in the position of the consumer in Adam Smith's classical market, freely choosing, based on fully transparent price and knowledge about quality, from a selection of products.

Nothing epitomizes this situation better than people with type 2 diabetes who require insulin, approximately 90 percent of whom are using hugely overpriced insulin analogs (discussed in chapter 4). The previous generation of insulin, recombinant human insulin, costs less than one-tenth as much and works as well as insulin analogs — sometimes even better — for the vast majority of people with T2DM. But when the drug companies are successful in creating the illusion of brand value, as they were with the insulin analogs for people with type 2 diabetes, they can charge prices commensurate with those exaggerated perceptions rather than with the true clinical value of the drug.

Probably the most extreme example of unregulated drug prices is Medicare Part D, which provides optional prescription drug coverage to seniors. The legislation, passed in 2003, explicitly prevents Medicare from using its purchasing power to negotiate drug prices with the manufacturers. This leaves Medicare paying 80 percent more than the Veterans Administration for exactly the same drugs.

Representative Billy Tauzin of Louisiana, former chairman of the

House Committee on Energy and Commerce, deserves special recognition in Pharma's Hall of Infamy for the role he played in writing that Medicare legislation and marshaling it through the House. Not long after passage of that bill, he announced he was retiring from Congress. He then took a spin through Washington's revolving door and landed in a $2 million-a-year job as head of the pharmaceutical industry's lobbying arm, the Pharmaceutical Research and Manufacturers of America (PhRMA). It was like a quarterback catching his own pass and then scoring a touchdown as a wide receiver.

The manufacturers defend their unregulated pricing by threatening that adoption of price regulation — which exists in all the other wealthy countries — would harm the public by choking off innovation. But this just doesn't jibe with the facts. In the United States, health technology is the most profitable sector of industry, earning a blue-ribbon 21.6 percent profit margin on sales in 2016. (Even though drug companies generated only one-fifth of the revenues in this sector, they raked in almost half the profits.) Calculated differently, from 2000 to 2018, the earnings as a percentage of sales were almost twice as high for the drug industry (13.8 percent) as for the S&P 500 (7.7 percent).

Neither does the so-called free-rider problem (the claim that other wealthy countries, all of which have lower prices, don't contribute their fair share of drug research and development costs) justify this level of profitability. In 2015, the manufacturers of the twenty top revenue-generating drugs in the world set their U.S. prices two and a half times higher than in other wealthy countries. The excess revenue collected by charging this premium in the United States ($116 billion) was half again greater than the manufacturers' *entire spending on global R&D* ($76 billion). In other words, these manufacturers could have paid for their entire R&D budgets from the higher prices charged in the United States and still kept $40 billion in excess profit. (Those inclined to defend Big Pharma's U.S. pricing might argue that this still doesn't prove that U.S. prices are not fair; but, as will be shown, the drug companies are drowning in so much surplus revenue, beyond what is needed to fund R&D, that they are investing massive amounts of money in stock buybacks and dividends.) So it's not true that the exorbitant drug prices in the United States are necessary to enable drug companies to continue to develop new drugs at the same pace.

To be fair, responsibility for the pricing problem does not belong to the manufacturers alone. Drugs (like Nexium, discussed in the previous chapter) that are on patent but don't offer meaningful advantage over much less expensive alternatives provide pharmaceutical benefits managers (PBMs, the middlemen between the manufacturers and consumer health plans) with the opportunity to extract large rebates from manufacturers. The PBMs collect the rebates in return for including the manufacturers' high-cost, low-value drugs in their formularies with low co-pays, thereby increasing their use. This kickback system creates a win-win for manufacturers and PBMs but extracts a high toll from the rest of society.

The cost of our prescription drugs is a huge problem in its own right. But an even greater problem is how our uniquely high prices (and profits) distort drug company incentives, which leads to commercial control of the knowledge that directs our health care.

MARKET FAILURE #2: ASYMMETRY OF INFORMATION

Price alone does not determine drug company revenues. As discussed in chapter 6, it's price multiplied by the volume of drugs sold, which is a reflection of doctors' belief that a given drug is in their patients' best interests. It's the manipulation of those beliefs that increases Americans' use of more expensive — but often *not* superior — new drugs.* And the main method of that manipulation is

---

* Among drugs discussed in earlier chapters, 88 percent of Neurontin, 80 percent of Vioxx, and 68 percent of Humira prescriptions were purchased in the United States.

to create a mismatch between, on the one hand, the totality of information available to the drug companies about their products and, on the other hand, the limited information they make available to doctors, insurers, policy makers, and the public.

Two primary tactics maintain this asymmetry. First, Pharma preserves an aura of normalcy around the lack of transparency of its scientific data and the consequent inability of experts to conduct independent analyses of the claimed clinical trial results prior to their publication. And second, unlike other wealthy countries, the United States prohibits any government agency from evaluating the comparative clinical benefit and fair price of new drugs to inform doctors, patients, and insurers about optimal use.

Meanwhile, doctors' reliance on the largely commercially controlled "scientific evidence" is so deeply ingrained that comprehending the extent to which this evidence has been hijacked would require, as Thomas Kuhn would say, living in a different world. And that's why doctors continue to tolerate the actual data from commercially sponsored clinical research remaining the proprietary intellectual property of the sponsor and therefore not available to the doctors or even to the experts who write the guidelines that direct medical care.

Government Consideration of Prescription Drug Cost-Effectiveness Is Prohibited

Rational decision-making about optimal prescription drug use requires that doctors, consumers, and insurers know both the clinical value (based on full access to clinical trial data) and the comparative cost-effectiveness of drugs. The United States has no mechanism for generating this information. The FDA evaluates each application for approval of a new drug (or a new use of an already approved drug) to determine whether the manufacturer's clinical trial data adequately demonstrate the drug is safe and effective for the requested indication. Despite being the only government agency that has access to the underlying clinical trial data, the FDA

- does not determine the clinical value of new drugs compared to the best available therapy;
- does not ensure that journal articles that publish the results of the clinical trials it has already reviewed are accurate and complete;
- does not correct published journal articles that fail on either count;
- and, finally, does not assess whether the price that will be charged for a new drug is fair in terms of the value it provides.

The FDA determines which drugs doctors *can* prescribe but provides little guidance about which drugs they *should* prescribe. Most developed countries — but not the United States — have agencies charged with determining the comparative value of new therapies to guide doctors' prescribing decisions as well as to inform national drug coverage and pricing policy. This process — called health technology assessment (HTA) — allows the market for prescription drugs in other developed countries to better serve society by at least partially overcoming the inherent advantage drug companies gain by control of their clinical trial data. Without HTA, our market for prescription drugs is about as orderly as professional sports would be without referees or umpires. Major League Baseball pitchers calling their own balls and strikes or NBA players calling their own fouls wouldn't work very well. When you are paid to win, competition is unlikely to be fair without impartial oversight.

To that end, the Institute of Medicine (IOM) issued a report in 2008 that proposed "a more systematic approach to evaluating [scientific] evidence for clinical effectiveness." The report recommended development of a U.S. government program to oversee production of unbiased systematic reviews of the medical literature and clinical practice guidelines to improve the quality of information available to doctors (although it did not call for transparency of clinical trial data). Notwithstanding the report's acknowledgment of the paradox of Americans' comparatively poor health despite the high cost of our health care, the sponsor, the Robert Wood Johnson Foundation, "urged the [IOM] committee to limit its work to the *non-cost issues* related to determining the effectiveness of health care services" (italics mine). Thus, the sponsor had tied the hands of the report's authors; a government agency to determine cost-effectiveness of new therapies was desperately needed, but the IOM was allowed to address only the efficacy side of the equation.

Then in 2010 came a stunning demonstration of Big Pharma's political muscle: The final version of the Obamacare legislation mandated continuation of the asymmetry of information by specifically *banning* consideration of cost-effectiveness studies in government-related coverage decisions or guideline recommendations. (As will be discussed in the following chapter, Big Pharma committed $150 million to pay for advertisements supporting passage of Obamacare.) A *New England Journal of Medicine* article asked: "How can our market-driven health system work efficiently if participants lack information about the relationship between the costs and benefits of health interventions?" The answer is simple: Our market-driven health system *cannot* possibly work efficiently when neither physicians nor patients know the value of new drugs compared to other therapies.

We now have several government organizations that evaluate new health-care products,* but each operates under a mandate *not to perform* full HTA activities that would use data on comparative cost to make practice and coverage recommendations. As discussed in chapter 4, if we had an HTA agency like those in the

---

* The FDA, as already discussed; the Agency for Healthcare Research and Quality, which promotes evidence-based practice but is not allowed to consider the cost of drugs in its recommendations; and the Patient-Centered Outcomes Research Institute (PCORI), which conducts research on the comparative effectiveness of drugs, again without being allowed to consider the costs of different therapies.

UK, Canada, Germany, and New Zealand, we would have saved more than $20 billion per year by relying on recombinant human insulin instead of the approximately 1,000 percent more expensive (but usually not clinically advantageous and sometimes even disadvantageous) insulin analogs for type 2 diabetes. And as discussed in chapter 6, Humira became by far the best-selling drug in the United States despite the fact that the manufacturer's own study showed that it was no more effective as first-line therapy for rheumatoid arthritis than methotrexate, which costs 99.5 percent less.

Evaluation of the comparative efficacy of therapeutic options typically must rely on nontransparent data, while the comparative cost of those options is specifically excluded from consideration. And as shown earlier, shortly after the Agency for Healthcare Research and Quality required a statement of compliance with "standards for trustworthiness" for each clinical practice guideline posted on its website, the program was summarily canceled. All of this leaves American doctors with little counterbalance to the commercially funded "surround-sound" marketing efforts supporting the use of expensive but not superior therapies.

Medical Journals' Complicity

Chapter 6 showed that reports of commercially sponsored clinical trials published in medical journals are significantly more likely to support the use of the sponsors' products. This raises the question of why respected medical journals allow their good names to be used to certify the results of clinical trials when the actual data have not been made available to editors or peer reviewers for independent analysis. Shortly after stepping down from his position as editor in chief of the *British Medical Journal*, Dr. Richard Smith wrote: "I must confess that it took me almost a quarter of a century editing for the *BMJ* to wake up to what was happening." What he woke up to was an awareness that the journals have a strong but largely covert incentive to publish articles that reflect favorably on brand-name drugs. The manufacturers, he realized, then purchase reprints of these articles, emblazoned with the journals' imprimatur, to serve as powerful marketing tools when handed out to physicians (like the *NEJM* article discussed in chapter 1 that falsely extolled the safety advantage of Vioxx).

Reprint sales are not just incidental to the primary business of the most prestigious journals. In 2005 they accounted for 41 percent of *The Lancet*'s total income. In 2012, eighty-eight "high reprint sales" articles published by *The Lancet* brought in an average of $455,000 each. Given an 80 percent profit margin on reprint sales, that's $31.7 million in profits in one year.

The *NEJM* refused to disclose its data on reprint sales, but Dr. Smith estimated its income was "likely to be even higher." In 2010 the *NEJM* led the world's top five journals in percentage of published trials that were manufacturer-supported, a whopping 78 percent. *NEJM* had the highest impact factor (the measure of the influence of a journal based on the frequency with which its articles are subsequently cited by other publications) and the highest percentage of industry trials, so it's a safe bet that *NEJM*'s reprint sales were also tops.

The sale of reprints by medical journals to the drug companies who sponsor the studies creates a major conflict of interest: The journals have an incentive to publish articles that report favorable findings so the manufacturers will purchase reprints. While authors of journal articles are required to declare conflicts of interest, there is no similar requirement for the journals themselves to provide data about the income they receive from reprint sales. The three most influential journals in the United States (*NEJM, JAMA*, and *Annals of Internal Medicine*) refuse to disclose this information, whereas the most influential journals in the UK (*The Lancet* and *BMJ*) do disclose it. Furthermore, only two of the twelve journals that are members of the International Committee of Medical

Journal Editors disclose financial ties between their editors and drug companies or other commercial entities.*

In 2015 Dr. Richard Horton, editor in chief of *The Lancet*, suggested an obvious remedy for the "turn toward darkness" in biomedical research. Dr. Horton pointed to two policies adopted by particle physicists after several "high-profile errors" in published research occurred in their field: "checking and rechecking of data *prior to publication*" (italics mine) and "filtering results through independent working groups . . . encouraged to criticise." Most readers, both medical professional and lay, assume that these safeguards have already been put in place by the powerful curators of our medical knowledge, *but they are wrong*.

The integrity of published reports of clinical trials would take a giant leap forward if medical journals simply required submission of the full clinical study reports (CSRs, described in chapter 6)† along with the manuscripts reporting the results of those trials. But journals have failed to take even this obvious first step necessary for peer reviewers and editors to ensure — *prior to publication* — that the important findings have been reported accurately and reasonably completely.

Requiring that CSRs be included with manuscript submission would involve little additional expense or effort. The CSRs would have already been produced by the drug company sponsoring the research, and redactions would rarely be necessary, except to protect patient confidentiality or truly proprietary information (like manufacturing processes). That this *could* be done with little extra cost is obvious. So why is it not being done?

---

* *BMJ* posts editors' conflicts of interest on its website; the *Annals of Internal Medicine* "discloses editorial COIs for certain types of online articles."

† The raw data from clinical trials are found in individual patient data (IPD) records, which are then tabulated and presented in CSRs. Thus, the IPD is more complete than CSRs but would require far more time, skill, and additional documentation from the manufacturer than reliance on CSRs for independent prepublication review of manuscripts.

We don't need to look beyond the effect that requiring inclusion of CSRs would have on the most prestigious medical journals. First, drug companies would have less incentive to publish study results in high-impact journals if the requirement for full prepublication data transparency constrained their freedom to spin and sometimes misrepresent clinical trial data. From the journals' perspective, the loss of these articles would compromise their prestige and reduce the impact-factor rating that determines advertising revenue. And second, reducing the potential for commercial bias through pre-publication data transparency would decrease the marketing value of the reprints to the manufacturers and thus decrease journal income from reprint sales. In other words, the journals' current incentives to publish influential articles that spawn profit-generating reprints are given precedence over the fulfillment of their responsibility to deliver the most accurate and complete scientific evidence to their readers.

Transparent *prepublication* peer review of clinical trial data is essential because once articles are published in respected journals and accepted by doctors as legitimate scientific evidence, the horses are out of the barn. Misrepresentations and material omissions discovered after publication are far less likely to have much impact on revising doctors' beliefs about the safety and efficacy of a drug. (Health-care economists refer to the reluctance of doctors to modify beliefs about the benefits and harms of drugs as "stickiness.")

A dramatic demonstration of this phenomenon occurred after the *NEJM* published its misleadingly incomplete article about Vioxx in November 2000. Just a few months later, the FDA posted Merck's data from the VIGOR trial, showing that Vioxx doubled the risk of serious cardiovascular events compared to an inexpensive over-the-counter pain remedy. Notwithstanding this clear and publicly available evidence of danger, the number of Vioxx prescriptions in the United States did not decline after this information became available on the FDA's website or even after the FDA Warning Letter to Merck (chapter 1) or the 2002 revision of the FDA-approved label for Vioxx. U.S. prescriptions for Vioxx continued at approximately twenty million annually from 2000, the drug's first full year on the market, until its withdrawal in 2004. And, again as discussed in the first chapter, the *NEJM* sold almost a million reprints of that article (mostly *after* it was aware of the unreported cardiovascular events), bringing in up to $800,000.

The obvious remedy for the medical journals' unwillingness to ensure the integrity of their clinical trial reports would be to collectively follow the example of the particle physicists described by Dr. Horton — require data transparency and independent analyses *prior to publication* of clinical trial results. The organization that could coordinate adoption of such a policy is the International Committee of Medical Journal Editors (ICMJE), composed of the editors of the most influential general medical journals in the world. But in 2016, rather than tackling the issue head-on by requiring submission of corroborating CSRs with manuscripts reporting clinical trial results, the ICMJE circulated a watered-down draft proposal. This would have required authors to agree to make individual patient data from clinical trials — but only the data relevant to published outcomes — available no later than six months *after* publication. Though adopting this proposal would have been a small step in the right direction, this limited postpublication transparency would have had no greater effect than did the public access to the FDA regulatory documents showing the increased cardiovascular risk of Vioxx, meaning not much.

Even so, Dr. Jeffrey Drazen, then editor in chief of the *NEJM,* coauthored an editorial expressing concern that, with access to this limited trial data, "a new class of research person will emerge," one who might "even use the data to try to disprove what the original investigators had posited." The editorial reported that "some researchers have characterized [such people] as 'research parasites.'" This from the man who was at the helm of the journal whose admittedly lax editing of the VIGOR trial report probably contributed to some thirty thousand American deaths. Can you imagine such "parasites" scrutinizing trial data to ensure the veracity of journal articles that influence the health and safety of millions? How dreadfully, despicably parasitic!

A year and a half later, a headline in *STAT News* read "New Science Data-Sharing Rules Are Two Scoops of Disappointment." The article reported that the final ICMJE "requirement for data sharing" was so diluted, it didn't even require data sharing, just a statement of data-sharing plans, which could consist of plans not to share the data. So the data-transparency can was kicked down the road by the medical journals without them even agreeing to limited postpublication transparency.

The journals' unwillingness to require that CSRs (and individual patient data when necessary) be submitted with manuscripts reporting their results stands as a victory for the drug companies and a clear example of market failure. The drug companies benefit from perpetuating the charade of peer review without data transparency, which allows them to continue exaggerating the advantages and minimizing the harms of their new drugs. The journals — especially the most prestigious — benefit by continuing to be the most-sought-after destination for reports of high-profile studies, which increases their impact factor and readership (and therefore advertising revenues) as well as their profits from reprint sales. And the commercially connected academic researchers benefit from the increased stature and financial rewards that accompany being associated with high-profile clinical trials.

All of which leaves hardworking doctors and their patients deprived of the full benefit of clinical trial evidence. But as long as the prestigious medical journals' primary consumers — practicing health-care professionals — are willing to turn a blind eye to the appallingly low standards of so-called scientific evidence in published articles, the journals have little incentive to adopt even the most basic standards of good science.

MARKET FAILURE #3: NEW MEDICAL KNOWLEDGE IS NOT DESIGNED TO ADDRESS AMERICANS' MOST PRESSING HEALTH NEEDS

The asymmetry of medical information is not limited to the integrity of the scientific evidence derived from commercially funded clinical trials; it also extends to what those trials are *about*. He who pays the piper calls the tune, and in the United States, private industry pays the piper for about five-sixths of clinical research. So it should come as no surprise that the kinds of things that are studied — and therefore become the knowledge that informs doctors' clinical decisions — are the kinds of things most likely to maximize financial return on research investment rather than to optimize Americans' health.

Out of the $121 billion the United States invests annually in medical and health services research (this includes both public and private funding), $116 billion (96 percent) is directed toward developing and testing new drugs and medical devices. The pittance that remains goes to studying how to deliver health care more effectively and efficiently and how to improve "the health and well-being of individuals, communities, and populations." Based on what almost all new health-related knowledge is *about*, doctors and their patients reasonably assume that relying on drugs, devices, and other biotechnological innovations — with a bias toward the newest therapies — is the best path to good health. The truth is that medical care accounts for no more than 20 percent of the variation in life expectancy in wealthy countries. Social circumstances and lifestyle factors are far more important. This leaves doctors — including really good doctors who are striving to practice evidence-based medicine — believing they are best serving their patients when they are often, in effect, acting as agents of Big Pharma.

The challenge of reducing the greatest health risk to people with type 2 diabetes — heart attacks and strokes — provides a current example. Trulicity is an injectable non-insulin drug for type 2 diabetes that was the sixth-best-selling drug in the United States in 2019. Although no more effective than metformin at lowering blood sugar, Trulicity costs up to 170 times more, $694 versus $4 per month. In February 2020, based on the results of the manufacturer-sponsored REWIND study, the FDA approved Trulicity for a second indication beyond just lowering blood sugar: "to reduce the risk of major adverse cardiovascular events in adults with type 2 diabetes mellitus who have established cardiovascular disease or multiple cardiovascular risk factors." The FDA's acknowledgment of cardiovascular risk reduction by Trulicity quickly became a major selling point.

A careful look at the REWIND study results published in *The Lancet* shows how commercially motivated medical research can manufacture "knowledge" that prioritizes drug company profits over health. The results — upon which the FDA's label change was based — show a statistically significant but minuscule reduction in risk of cardiovascular events associated with Trulicity, making the cost per event prevented enormous: 323 people must be treated with Trulicity for one year to prevent one cardiovascular event, without significantly reducing the overall risk of death. At the drug's current price, the cost of preventing a single nonfatal heart attack or stroke with Trulicity in the United States is almost $2.7 million.*

For this small benefit to reach statistical significance, the REWIND study, funded by Eli Lilly, required almost 10,000 people to be followed in 374 clinical sites located in 24 countries for 5.4 years. Nevertheless, the FDA approval resulting from this enormous outlay paid off handsomely, allowing Lilly to advertise that Trulicity reduces cardiovascular risk in people with T2DM. But, you may ask, how does this tiny advantage compare to far less expensive programs encouraging people with T2DM to adopt and maintain healthy lifestyle habits? We simply don't know because neither Lilly nor any other drug company has been willing to fund that study.

---

* Trulicity costs $8,328 per year, and 323 people must be treated to prevent one event: $8,328 x 323 = $2,690,000 per event prevented.

And that's exactly the point: The purpose of U.S. medical research is now to maximize financial returns generated by innovation in biotechnology, whereas the evidence for many of our most prevalent diseases strongly suggests the far greater benefit of "upstream approaches," such as lifestyle modification, to reduce the burden of disease. Left to its own devices, the market has no incentive (actually, it has a negative incentive) to fund research to determine the interventions with the greatest *health* benefit.

Allowing our medical research agenda to be driven by the drug industry's short-term financial interests definitely means that we get more new therapies. But it also means that the medical knowledge doctors rely on will be about what is important to the sponsors of clinical trials rather than what would be most helpful to the American people.

MARKET FAILURE #4: EXCESSIVE RETURN ON BIOTECH INVESTMENT

Investors — whether institutions (like venture capital or private equity firms) or individuals — seek to maximize their return on invested capital (ROIC).* To that end, the synergy between unregulated monopolistic pricing of brand-name drugs; commercial control over the research, and communication of the results of that research to doctors through the sources they trust the most; and the lack of a government agency with access to *all* clinical trial data to determine the true health and economic value of new therapies has turned Big Pharma into investors' ultimate golden goose.

Profit margins for Big Pharma are consistently at least twice as high as those of the other five hundred largest global corporations,

---

* ROIC equals net operating profits after taxes divided by invested capital (book value of company stock, interest in subsidiaries, and outstanding long-term debt). ROIC correlates strongly with the changes in stock price; see forbes.com/sites/greatspeculations/2018/11/27/dont-get-misled-by-return-on-equity-roe/#7495d6fe4ed4.

which translates into by far the highest ROIC for biotech among all industries. Between 2011 and 2019, biotech investors earned an average of 17.3 percent annually. The next-highest-earning industry, lodging and food services, returned 2 percent less, and the average for all industries (except Big Pharma) was 11.5 percent.

Big Pharma's consistently outsized ROIC ultimately comes from working Americans who pay for their products through the cost of their insurance, suppression of their wages, consumption of public resources, and cash (for co-pays and, for those who lack insurance coverage, payment in full). As we noted, most Americans are upset about the price of drugs, but investors need not fear: The fate of a bill passed in 2019 by the Democrat-controlled U.S. House of Representatives illustrates how difficult it has been and will continue to be to rein in those excessive returns. The bill called for limiting the price paid by Medicare, Medicaid, and private insurance for the highest-revenue-generating prescription drugs. It would have required price negotiation with the manufacturers of at least the 50 (and up to a total of 125) drugs Americans spend the most on, aiming at a price ceiling of no more than 120 percent of the average paid in six European countries. Predictably, PhRMA was up in arms, hyperbolically warning the bill would cause a "nuclear winter" in biotech innovation.

The Congressional Budget Office (CBO) came to a very different conclusion. They calculated that, had this bill been approved by the Senate and signed by the president, it would have reduced Big Pharma's revenues by $456 billion for the decade starting in 2020. However, said the CBO, a reduction in Pharma revenues of this magnitude would likely mean that the total number of new drugs approved over the next ten years (estimated at three hundred) would be reduced by only "approximately 8." And since only one out of eight new drugs provides previously unavailable benefit, this would mean that only one true advance in therapy would be missed over ten years.

But let's look further at the $456 billion that would not have been extracted by the drug industry if the House's bill had been enacted.

Without that revenue, the biotech industry's ROIC still would have remained more than a percentage point higher than that of the next-highest industry. Moreover, this calculation assumes no reduction in the price the large drug companies pay to acquire the rights to drugs in development or purchase early-stage research companies. Unless the pharmaceutical industry insists on maintaining its 2 percent higher rate of return on investment than any other industry, innovation would not need to be compromised at all. Even a one-*trillion*-dollar reduction in drug company revenue over the coming decade would still leave the pharmaceutical industry with the highest ROIC.

However, these issues were rendered moot when Mitch McConnell, then majority leader of the Republican-controlled Senate, refused to introduce the legislation for consideration. He rejected negotiation of drug prices at the federal level as "socialist price controls."

To protect itself from the growing public anger and increased media coverage about the high prices of prescription drugs, PhRMA committed many millions of dollars to its multiyear "Go Boldly" media campaign, attempting to keep the public in the thrall of biotech innovation: "America's Biopharmaceutical Companies go boldly into the search for new treatments and cures, everyday [*sic*]. They are pioneers in innovation, ushering in a new era of treatments for patients." Their message was finely honed to engage Americans' hopes for cures of cancer, Alzheimer's disease, and many of our other frightening vulnerabilities. And the campaign skillfully obscured the fact that most newly approved drugs — despite being labeled *innovative* — do *not* provide previously unavailable benefit.

PhRMA's PR campaign against the bill to limit drug prices in the United States shows just how shameless drug companies are willing to be to protect their excess profits. The $456 billion that the bill would save over ten years — ostensibly causing a "nuclear winter" in drug innovation — was dwarfed by the $577 billion the drug industry had awarded its investors in the form of stock buybacks and dividends in just the five years between 2016 and 2020. To be clear,

PhRMA was threatening the public with dire health consequences if legislation brought exorbitant drug prices back in line with prices in other wealthy countries, yet at the same time its companies were skimming off more than twice as much for their investors! This is disingenuousness and greed in the first degree, yet PhRMA's clever media campaigns continue to succeed in confusing the public.

In her book *These Truths: A History of the United States,* Harvard professor of American history Jill Lepore explains that innovation emerged as a business concept that was not so much about providing social benefit as about introducing new products to replace older products: "Innovation might make the world a better place, or it might not; the point was, innovation was not concerned with goodness; it was concerned with novelty, speed, and profit." Though rarely stated in polite company, the phrase "novelty, speed, and profit" describes the primary purpose of biomedical innovation perfectly: to persuade doctors to replace current therapies with maximally *profitable* new therapies, irrespective of actual medical or economic value.

That said, some new drugs provide tremendous benefit but are expensive and must be used for the patient's lifetime. Prominent among these are the drugs that transformed HIV infection from a death sentence to a chronic condition compatible with a normal life span. In the United States, the cost of Atripla, one of the most frequently used combination pills to treat HIV, is over $35,000 a year. In India, which does not have patent-protected monopoly pricing, the generic version of the same drug costs about $100 per year.

Some drugs are highly effective and only used short-term, but are outrageously priced, like those that treat hepatitis C. Gilead purchased the rights to the first of these drugs, Sovaldi, for $11 billion from a small biotech company, Pharmasset, whose research had been largely funded by the federal government. The value of Gilead's acquisition was based on projections of an eighty-four-day course of therapy costing $36,000. But Gilead immediately more than doubled the price, to $84,000, and recovered the entire cost of its acquisition in the first year of sales. Gilead's price was certainly not justified by the cost to manufacture Sovaldi — the company licensed the right to make biosimilar (similar to generic) Sovaldi in India for as little as $4.29 per pill, not the $1,000 per pill it cost in the United States. But equally certainly, Gilead's price provided a windfall for investors: Between 2016 and 2018 Gilead was the most profitable large corporation in the United States, logging 49.7 percent profit on its sales.

Some new drugs provide significant albeit limited relief for serious life-long conditions but are unreasonably priced, like the drugs that partially and temporarily correct the genetic defect responsible for cystic fibrosis. These drugs improve lung function between 3 and 14 percent, improve quality of life, and decrease the frequency of pulmonary infections, but they have not yet been shown to improve survival rates. Even so, they carry the "miracle drug" price of about $300,000 per year.

The chief medical officer of the nonprofit Institute for Clinical and Economic Review (ICER) in the United States summarized his organization's report on the latest of the cystic fibrosis drugs: "Despite being transformative therapies, the prices set by the manufacturer — costing many millions of dollars over the lifetime of an average patient — are out of proportion to their substantial benefits." ICER estimates that the price of the most recent drug to treat cystic fibrosis, Trikafta (Vertex), would need to be reduced by about three-quarters for the drug to be in the reasonably cost-effective range. (Lest the ICER report be interpreted as an anti-industry hit job, two of the three expert reviewers of the report had financial ties to the manufacturer, each having received more than $40,000 in 2018.) This price, however, has worked out very well for investors: The value of Vertex stock increased sixfold during the eight years starting in April 2012.

Here's the kicker: Americans are so mesmerized by biotechnological innovation that we fail to see the less expensive and more effective approaches that are right in front of us: A study published in 2017 showed that Canadians with cystic fibrosis live an average of *ten years longer* than Americans with that disease (50.9 years versus

40.6 years). Americans have greater access to expensive new thera-pies, but Canadians are better at attending to the basics of care, with nutritional support playing a key role. And, paradoxically, Canadi-ans with severe cystic fibrosis are 50 percent more likely to receive lung transplants than are Americans with equally severe disease. One variable that does matter is insurance coverage — Canadians have universal insurance. Americans with CF who have good insur-ance do as well as Canadians.

Is our top priority decreasing the burden of disease in people with cystic fibrosis, or is it implementing new technology? If it's the former, the first step would be to learn from the Canadians' success. It is possible that a combination of the Canadian approach to cystic fibrosis care *plus reasonably priced new drug therapies* would pro-duce the best outcomes, but making that determination requires an investment in research that would not necessarily benefit the bio-tech industry.

Many of the highest-revenue-generating drugs in the United States simply do not add clinical benefit and do not even appear on the World Health Organization's list of essential medicines. Some of these were discussed in earlier chapters — insulin analogs to treat type 2 diabetes (chapter 4),* biologic therapies like Humira as first-line drugs to treat rheumatoid arthritis (chapter 6), and Nexium to block stomach acid (chapter 8). And, worse, some new drugs and devices do more harm than good, like the long-acting opioid Oxy-Contin and metal-on-metal hip implants.

Irrespective of whether these drugs are cost-effective for Amer-ican patients, they're certainly profitable for investors. With the higher cost and greater use of expensive new drugs in the United States, conservatively two-thirds and more likely three-quarters of global drug company profits are "earned" in our country. This leads to overcapitalization of the pharmaceutical industry, which then funds a new round of product development and marketing, which

---

* With this exception: a weak recommendation for long-acting insulin ana-log in individuals with "frequent severe hypoglycemia with human insulin."

ever more skillfully shapes the beliefs of doctors and the public, which increases the use of innovative, overpriced, but mostly *not* clinically superior and rarely cost-effective products. A small portion of these profits is reinvested in lobbying, campaign contributions, and public relations, all of which help keep drug-price controls and health technology assessment at bay. And the cycle repeats.

THE CO-OPTED NARRATIVE OF AMERICAN HEALTH

The market failures described in this chapter serve the financial interests of the drug companies' executives and investors, medical journals, academic medical centers and researchers, physicians' professional societies, and health-related nonprofit organizations. The scientific evidence produced and disseminated by this nexus of intertwined interests imposes a single unified narrative: Biomedical innovation is Americans' primary path to better health. Doctors have little choice; their practice would not conform with community standards if they failed to base patient-care decisions on these sources of information.

For the public, incessant drug ads and well-funded public relations campaigns on television, in print, and on social media reinforce the narrative of the centrality of technological innovation to improved health. These messages resonate with people's hopes for easy fixes for their health challenges, leading Americans to believe that 80 percent of our increased longevity is due to improvements in medical care. But this faith in the efficacy of medical therapy is unrealistic.

Doctors' dependence on scientific evidence from sources that exaggerate the importance of biotech innovation works synergistically with both the public's unrealistic expectations of the efficacy of new medical therapies and our lack of constraint on medical spending, producing a predictable result: We now spend about

65 percent more of our GDP on health care than the average percentage of eleven other wealthy countries.

Our greater reliance on the profit-maximizing medical knowledge generated by this market-driven system explains half of the paradox of American health care — why our spending is so far out of line. But it does not explain why, in the face of so much more spending, Americans' health is not at least equal to that of the citizens of other wealthy but lower-spending countries. Here's the key to the second half of this paradox: Whereas the market-driven narrative that exaggerates the benefits of medical innovation is perfectly rational from a *commercial* point of view, it is upside down from an *epidemiological* point of view.

The factors that influence the way we live our lives and the opportunities we have play a far greater role in our likelihood of remaining healthy than does the medical care we receive (although that too is important). Socioeconomic factors like education, employment, income, family and social support, and community safety are responsible for 40 percent of Americans' health status; health behaviors like tobacco use, diet and exercise habits, alcohol use, and unsafe sex determine 30 percent; and environmental quality and the built environment (that is, infrastructure) account for another 10 percent. Thus, medical care determines not 80 but 20 percent of our health. Yet only 4 percent of medical research investigates how to deliver more effective and efficient health care because the knowledge produced by such research offers little, or even negative, monetary benefit to those who fund it.

Our national delusion is accepting this commercially directed body of knowledge as if it were a *public good,* whose purpose is to improve the health of all Americans. Rather, this knowledge is produced and disseminated because of its value as a *private good,* the purpose of which is to benefit the commercial interests of the sponsors, investors, researchers, and outlets that deliver that information (including medical journals). Our national acceptance of the so-called knowledge that is produced and disseminated with the primary purpose of maximizing return on research investments is reflected in our acceptance of committing less than 2 percent of our national health expenditures on population-based disease prevention. These chickens came home to roost during the early phases of the COVID-19 pandemic, when we failed to use low-cost public health measures (masking, social distancing, handwashing, testing, and contact tracing) and instead turned to unproven medicines (like hydroxychloroquine and remdesivir) to rescue those who became ill.

The final piece of this puzzle is how comparatively little we spend on the social factors that determine the other four-fifths of our health. In *The American Healthcare Paradox: Why Spending More Is Getting Us Less,* Elizabeth Bradley and Lauren Taylor showed that, compared to other OECD countries, we spend a far greater percentage of our GDP on health care but a far lower percentage than most on "social services" (such as supports for older adults, disability benefits, rent subsidies, and so on). For every dollar the other OECD countries spend on health care, they spend two dollars on social services. In contrast, for every dollar we spend on health care, we spend less than sixty cents on social services.

Furthermore, it's not just our comparatively low spending on the non-medical-care determinants that compromises our health. Bradley and Taylor found that the higher a nation's *ratio* of health care to social-service spending (regardless of the absolute amount), the *lower* the life expectancy and the *higher* the rate of infant mortality. Excess commercially directed spending on medical innovation, it turns out, not only drains our national resources but, due to its primacy over addressing the upstream determinants of health, is actually harmful.

Of course, the drug companies aren't responsible for all that's wrong* with American health care or Americans' health. Still, the

---

* Nonpharmaceutical categories of unproductive spending include unnecessary administrative complexity (especially billing), failures in care delivery (especially failure to provide preventive care), overtreatment (especially at the end of life), and fraud and abuse.

negative consequences of the pharmaceutical industry's business model reach far beyond the extra $170 billion they take in each year through unregulated and unjustified pricing.

As I pointed out in the introduction, American health care has devolved into a "tail wags dog" situation. The invisible hand of the market, even the health-care market, has no social conscience. Nor does it have any commitment to the principle that all Americans deserve the opportunity to reach their health and life potential. Rather, our innovation-based, commercially funded, ROIC-maximizing narrative about health care sustains the appearance of normalcy surrounding our upside-down distribution of resources between medical care and the social aspects of health.

Which brings us back to the conundrum I had inadvertently confronted the pension fund managers with. By maximizing financial returns for the public employees whose pensions they oversaw, they were fulfilling their fiduciary responsibility. But in doing so, they were also strengthening this system of excess profitability that allows Pharma to maintain its influence over doctors, the public, and our political system, influence that is contrary to the employees' broader interests. And don't hold your breath waiting for the big drug companies to have an epiphany of social conscience that leads them to voluntarily cede their power and profitability in the service of the greater good. *It's not their job.*

Therein lies the core challenge. Our health care has become so subjugated to the values of the market that Big Pharma can mislead our doctors, misdirect our health care, and contribute to the misallocation of our nation's resources. Meanwhile, the political and economic strength of the drug companies and other health-related industries leaves us with bleak prospects for rebalancing American health care and population health.

The resilience of our dysfunctional health-care system becomes even more painfully obvious when we consider the fate of Haven, a joint venture formed by three major American corporations — Amazon, Berkshire Hathaway, and JPMorgan Chase. Haven had been created in 2018 with the goal of pooling the resources and technological prowess of these three companies to disrupt health care "business as usual" in order to bring more effective and efficient care to employees and their family members, 1.2 million in number. The prospect of this joint venture was so threatening to health care's vested interests, the mere announcement of it caused the loss of billions of dollars in the value of their stock. But Haven failed.

In January 2021, *STAT News* reported: "The dissolution of Haven marks one of the most stunning collapses in modern health care history." What caused this failure? Several theories have been offered, including the complexity of American health care and the lack of managerial experience among Haven's otherwise highly respected leadership. But a more likely explanation is that all three of the corporations involved in the joint venture had significant financial interests in U.S. health care that were in direct conflict with the goals of Haven. Berkshire Hathaway held investments in several large drug companies as well as in the largest U.S. provider of kidney dialysis. Amazon was getting into the mail-order pharmacy business and expanding services to health-care businesses and hospitals through Amazon Web Services. And JPMorgan Chase offered a broad array of banking services to the health-care industry: "From biotechnology to pharmaceuticals, medical devices to diagnostics—as you make history, we'll be with you every step of the way."

The failure of Haven shows that effective reform will require a broad coalition comprising:

- health-care professionals demanding accurate, complete, and epidemiologically balanced information to guide patient care;
- health-care purchasers—businesses both large and small whose financial success does not depend on profits from health care, as well as government and unions—seeking the most cost-effective care; and
- health-care consumers seeking the best health possible.

In other words, all of us who believe that the fundamental purpose of American health care should be maximizing health most cost effectively rather than maximizing the profits of the health-care industry.

Part I of this book showed how the clinical benefits of four best-selling drugs and classes of drugs had been exaggerated to doctors, leading them to provide care that was unnecessarily expensive and not in their patients' best interests. Part II has described how the radical narrowing of the core mission of American corporations has contributed to the undermining of the relevance and integrity of the sources of medical information that doctors trust. Although doctors may believe they have no alternative — and many are not even aware they should be looking for one — there is a path toward correcting the commercial bias in our medical knowledge and rebalancing our national investment in medical care and the social determinants of health. This is addressed in part III.

PART III

Moving Forward

9

THE LIMITS OF OBAMACARE

Politically and emotionally, I would have found it a lot more satisfying to just go after the drug and insurance companies and see if we could beat them into submission. They were wildly unpopular with voters, and for good reason. But, as a practical matter, it was hard to argue with [Senate Finance Committee Chairman] Baucus's more conciliatory approach.

— BARACK OBAMA

The U.S. health-care industry's "business as usual" has pulled major portions of American health care into what Harvard professor of government Michael Sandel describes generally as the domain of "market society," the purpose of which is to maximize financial gain rather than social benefit. Without adequate oversight of the drug companies' business practices, the public's benefit from medical science is too easily sacrificed to industry's financial goals.

Given our relentless progression over the past forty years toward extremely high-cost health care and the comparative decline in our population health, Americans will surely continue to demand reform. But we will not achieve meaningful and sustained progress on the most urgent issues, like improving the health of all Americans, instituting universal health-care coverage, and controlling health-care costs, until we correct the commercial distortion of the scientific evidence that informs our health care.

Which is not to say there haven't been full-bore efforts at reform, including the unsuccessful Clinton health plan put forth in 1993 and, more recently, Obamacare, which is the primary focus of this chapter. The success of Obamacare in extending health-care coverage to many millions of Americans is clear. But in the legislative process that produced Obamacare, broader reforms — including government-sponsored cost-effectiveness research and a public option to compete on the exchanges with private insurance plans — met with insurmountable resistance. Any future reforms will likely encounter as much, if not more, pushback.

OBAMACARE: ACCOMPLISHMENTS AND LIMITATIONS

The current era of U.S. health-care reform began in 2010 with the passage of the Patient Protection and Affordable Care Act, aka Obamacare. The final product of the legislative battles provides important lessons about the limits of reform, even in the context of Democratic control of both houses of Congress and the presidency.

During the 2008 presidential campaign (Obama versus McCain), Senators Max Baucus (D-Montana) and Charles Grassley (R-Iowa), the leaders of the Senate Finance Committee, convened a bilateral "Health Reform Summit" to explore the opportunities and impediments the next administration would face in its attempts to repair "our broken health care system." Federal Reserve chairman Ben Bernanke opened the summit with keynote remarks that focused on the current and future negative impact of rising health-care costs on the American economy.

Soon thereafter, Senator Baucus offered the following observation to Chairman Bernanke: "Some people suggest that because the American system is so complex and because it is so tied to the political pressures in all different segments, whether it be pharmaceuticals [or] insurance companies . . . that perhaps we should look at some kind of a Federal health board, somewhat patterned after

the Federal Reserve system, to help solve some of these problems."
Chairman Bernanke responded that a federal board might be helpful in the same way that an independent commission had served
to insulate legislators from political blowback following closure of
military bases in their districts. Although Senators Grassley and
Mike Crapo (R-Idaho) also supported the idea, increased federal
oversight of the health-care industry went no further.

Economist Peter Orszag, then head of the nonpartisan Congressional Budget Office, told the summit that the United States was
wasting $750 billion annually on excess health-care spending. But,
explained Orszag, cost-saving reform was inherently difficult because the money would be taken away from powerful and concentrated vested interests that exercised political influence through
lobbyists, campaign contributions, and political messaging. Orszag's
lesson in realpolitik directed the focus of reform away from trying
to cut costs as health care was expanded.

After the 2008 election, health-care economist Jonathan Gruber reinforced the takeaways from the summit to the Obama transition team: "You can either try to expand coverage or you can try
to do something to control costs. But trying to control costs too
much dooms whatever you do, because the lobbyists will kill you."
As Gruber had found in his work on Romneycare (health-care reform implemented in Massachusetts in 2006 that relied heavily on
recommendations from the conservative Heritage Foundation and
served as the template for Obamacare), the health-care industries
will support expansion of coverage "because that creates more customers."

Notwithstanding this advice, once in office, President Obama
struggled to achieve bilateral agreement on a plan that would both
expand coverage *and* control costs. As described by Steven Brill in
America's Bitter Pill, Obama understood that the plan's public option* — potentially the most powerful cost-containing feature of

---

* Obamacare initially included a public insurance option to compete with
private health insurance coverage in the hope of driving down both premi-

this phase of reform—came in a wide range of variations. At one end of the spectrum, a public plan would pay doctors and hospitals at the same rates paid by Medicare, which, when combined with the administrative savings compared to private insurance, would lead to significantly lower premiums. At the other end, a public option would be required to forgo these pricing advantages, which would shrink the difference between the public option and private insurance offered on the exchanges, rendering the exercise essentially meaningless. Still, as Brill reported, "the insurance companies, hospitals, doctors, drugmakers, and every other industry sector opposed even the seemingly weaker version [of a public option], as did Republicans and other free market supporters." The vested interests banded together to keep the camel's nose out of the tent of health-care reform.

On March 5, 2009, President Obama convened a daylong health-care-reform meeting at the White House that included 150 legislators and many stakeholders, such as CEOs, union leaders, health-care professionals, and consumer representatives. The president pressed the issue: "If we don't address costs, we will not get this done. If people think we're simply gonna take everyone who's not insured and load them up into a system . . . the federal government will be bankrupt. State governments will be bankrupt. I'm talking to you liberal bleeding hearts out there. . . . Don't think that we can solve this problem without tackling costs."

Although these comments were ostensibly addressed to the "liberal bleeding hearts" in the audience, I suspect President Obama was also telling the representatives from the drug, insurance, and hospital industries in the room that they too would have to participate in cost-cutting reforms.

But as it turned out, President Obama was the one who would ul-

---

ums and underlying health-care costs. Progressives supported a public option as competition to private insurance and possibly a transition toward single-payer insurance, while conservatives opposed it as a government "takeover" of health care.

timately have to yield; late in the process, the two primary cost-cutting strategies were removed from the legislation. The first involved the plan's Patient-Centered Outcomes Research Institute (PCORI), which had been introduced to "help people make informed health-care decisions." But as the legislation evolved, the institute was specifically prohibited from performing cost-effectiveness studies that would "establish what type"* of health care is cost effective or recommended. In other words, Obamacare was forbidden to make exactly the kind of comparative analyses that were necessary for the market to create efficiency.

The second cost-saving reform sacrificed to the legislative process was the public option. This was rejected by conservative Democrats, especially Senator Joe Lieberman of Connecticut, whose capital city, Hartford, is nicknamed "the insurance capital of the world." President Obama wrote in his memoir that after removal of the public option, necessary to nail down the yea votes of a filibuster-proof sixty senators, "activists on the left went ballistic." He continued, "I groused to my staff, 'Should I tell the thirty million people who can't get covered that they're going to have to wait another ten years because we can't get them a public option?'"

The primary thrust of the final legislation was providing government support to help poor and lower-income Americans obtain health-care coverage they could not otherwise afford and to prevent insurers from discriminating against people with preexisting conditions. The health-care industries could squeeze no money out of the poor (those making less than 138 percent of the federal poverty level), so for them, full coverage was offered through Medicaid expansion — except in twelve states, where the ideological commitment to limit government spending† still exceeded the moral com-

---

* Cost-effectiveness studies would inform health-care providers about the most effective *and* efficient therapies. For example, such analyses, if available, would have encouraged doctors to start people with type 2 diabetes mellitus on far less expensive but at least equally effective recombinant human insulin instead of the newer insulin analogs.

† The federal government initially covered 100 percent of the cost of

pulsion to provide health-care coverage to those least able to afford it. For nonwealthy Americans with incomes above this level, insurance was partially subsidized to bring the premiums within reach, which then pulled more family resources into the health-care system and created more retail consumers.*

The pharmaceutical industry was so happy with the guaranteed influx of non-cost-controlled retail customers, it pledged $150 million to pay for advertisements in support of Obamacare. AHIP, the lobbying arm of American Health Insurance Plans, supported the plan because, without Obamacare, states' movement toward mandatory coverage of preexisting conditions was pushing the cost of non-group insurance so high that this niche of their market was at risk of collapsing. In the four years following the passage of Obamacare, the value of the stock of large health insurers doubled, going up twice as much as the Dow Jones Industrial Average.

When the dust finally settled, the reforms included in Obamacare remained safely within the realm of win-win in the sense that more uninsured Americans would gain coverage and more money would be made by the drug, insurance, and hospital companies. And this would hold as long as taxpayers were willing to foot the bill for those subsidies.

By 2016, Obamacare had decreased the number of uninsured Americans by 40 percent, to a low of 26.7 million — an improvement, for sure, yet still a far cry from universal coverage. But it did little to curtail health-care costs or improve the quality and relevance of the scientific evidence available to doctors and other health-care professionals.

Obamacare's Medicaid expansion, meaning there was no cost to the states; that amount declined gradually to 90 percent by 2020.

* Of course, expanding coverage is a good thing. Here, I am not addressing this from the moral or medical perspective but from industry's business perspective.

IGNORING THE WILL OF THE PEOPLE

The failure of Obamacare to address many popularly held views about health-care reform was not an anomaly. In a now-classic 2014 paper, Professors Martin Gilens and Benjamin Page presented their analysis of a massive data set to quantify the impact of different groups on U.S. government policy. They came to an unfortunate but not unexpected conclusion: "Economic elites and organized groups representing business interests" not only exert substantial impact, but their policy positions tend to be *exactly the opposite* of those of non-influential ordinary citizens.

American health care is probably the best example. Almost nine out of ten Americans (87 percent) favor congressional action to lower the cost of prescription drugs. Nonetheless, after a bill to contain drug costs (discussed in chapter 8) was passed by the Democrat-controlled House in December 2019, the Republican Senate majority leader, Mitch McConnell — who received more money from Pharma than any other legislator in the first eight months of 2020 — refused to even bring it to the floor for consideration. The drug companies' pricing practices emerged unscathed.

Other issues also prove the point. Eight out of ten Americans (79 percent) are dissatisfied with the cost of their health care. Seven out of ten Americans believe that health care should be a guaranteed right for all. And almost seven out of ten Americans (68 percent) favor adding a public option to compete with commercial insurance. But our government, reflecting the strong influence of Big Pharma and other "economic elites," has not meaningfully addressed any of these publicly supported positions since the passage of Obamacare.

Instead, Pharma once again demonstrated its influence by gaining overwhelming legislative support for the 21st Century Cures Act in 2016. The bill, actively supported by thirteen hundred lobbyists, most of whom were paid by the drug companies, added $6.3 billion to the search for new therapies for cancer and Alzheimer's disease and ways to advance genomics-based "precision medicine."

What could be wrong with additional funding for medical research?

Two things. First, just beneath the veneer of accelerating medical progress, the bill included a provision to weaken the quality of the scientific evidence necessary to demonstrate the safety and efficacy of new drugs and devices. Formerly, approval of a new drug required supporting data from randomized controlled trials, the gold standard for evidence-based medicine. The new act lowered the bar to allow evidence of drug safety and effectiveness from observational (non-controlled) trials and even from collected reports from individual patient experiences to be used by the FDA in support of drug approval.

In addition, indirect evidence of clinical benefit, called surrogate end points (such as lowering cholesterol levels or reduction of amyloid plaque in Alzheimer's patients), instead of bona fide clinical outcomes or end points (such as reduction in cardiovascular events or maintenance of cognitive function), could now be accepted as end points in clinical trials. Supporters of the bill claimed that lowering the evidentiary bar would help correct the slow pace at which the FDA approved potentially helpful new therapies. But the FDA's alleged slow pace was fiction. The truth is that the FDA already provides the most rapid drug approvals in the world, and lowering the bar for scientific evidence is likely to do more harm than good for patients (though not for the drug companies).

And second, more than half of the funding for the bill, $3.5 billion, was *taken away* from money in the Obamacare Prevention and Public Health Fund that had been earmarked specifically for underfunded population-health interventions, like vaccination and smoking-cessation programs, as well as for "detecting and responding to infectious diseases and other public health threats," like the COVID-19 pandemic that would soon arrive. Because of this, the leaders of the American Academy of Family Physicians and the American Public Health Association opposed the act. Speaking as one of only five senators who voted against the bill, Senator Elizabeth Warren (D-Massachusetts) railed: "When

American voters say Congress is owned by big companies, this bill is exactly what they are talking about."

Passage of the 21st Century Cures Act shows the extent of bipartisan legislative support that Big Pharma and medical-device makers enjoy. In 2020, more than two-thirds of senators and representatives — split almost evenly between Democrats (47 percent) and Republicans (53 percent) — accepted money from the pharmaceutical industry.

Obamacare reduced the percentage of uninsured U.S. adults from 20 percent in 2010 to 12 percent in 2018. But expanding coverage required giving up other goals; there were virtually no cost-saving measures or cost-effectiveness analyses to determine the most effective *and* efficient approaches to improving health, and there was no public option. And paradoxically, as the percentage of uninsured Americans declined, U.S. longevity not only failed to improve but actually *decreased*. Even without these industry-threatening measures, Obamacare never enjoyed bipartisan support, and the Republicans continued their attempts to repeal it or have it struck down by the courts.

Going forward, reform will continue, at least initially, in the same direction as Obamacare, with the goal of further reducing the number of uninsured Americans. The proposed expansion of coverage will be achieved by increasing federal subsidies to make insurance affordable for more lower- and middle-income Americans. But, as President Obama made clear early in the legislative negotiations that culminated in Obamacare, the goals of reform will not be met without *both* expanding coverage and containing costs. And this cannot be done without proceeding into the territory that economists Orszag and Gruber warned would "doom" reform efforts. This is the subject of the next chapter.

10

THE KEY TO MEANINGFUL REFORM: FIX THE KNOWLEDGE PROBLEM

> Corporations have a valuable role to play in American society, and they contribute primarily by trying to make money.... The missing part is the role of government in ensuring that those profits do not come at the expense of society.
>
> — BINYAMIN APPELBAUM, *New York Times*

Obamacare extended health-care coverage to 40 percent of previously uninsured Americans and the profits of the drugmakers, insurance companies, and hospitals increased, but, as discussed in the previous chapter, the overall health of Americans declined. The health-care reform we need now — to improve Americans' health, reduce health-care costs, and move toward universal coverage — must break out of the politically safe confines of "win-win," in which coverage is expanded only by increasing the profits of the vested interests. This will not get us where we need to go. We must enter the realm of zero-sum reform, where private interests are no longer allowed to maximize their profits at the public's expense, but rather public and private interests are realigned to better serve society as a whole.

With regard to Big Pharma, this will require rebalancing the market for prescription drugs so that manufacturers compete based on the true effectiveness, safety, and value of their products (as determined by independent analyses of all clinical trial data) rather than on the manufacturers' often unsubstantiated claims. This chapter outlines the three steps necessary to achieve the reforms that would rein in the largely unrestrained power of Big Pharma and fix the underlying knowledge problem.

STEP 1: ADDRESS THE IMPEDIMENTS TO UNIVERSAL COVERAGE

Providing health-care coverage for all Americans won't be possible until the cost of U.S. health care is brought back in line with that of the other wealthy nations, all of which provide at least near-universal coverage, have far healthier populations, and spend far less on health care. The containment of health-care spending necessary to achieve near-universal coverage in the United States would throw a big wrench into Pharma's precisely tuned mechanism for maximizing the profits they currently reap from U.S. drug sales. This is why the drugmakers unite to support expansion of coverage to Americans who would otherwise not be able to afford health care (and thus do not currently contribute to Pharma's profits) but remain adamantly opposed to expansion of coverage with inherent cost-saving mechanisms to Americans who currently participate in the health-care system as retail customers through private insurance. From this perspective, Pharma's response to proposals that would expand coverage and offer cost-saving options to their current retail customers is entirely predictable.

In 2020, then-candidate Joe Biden revived the debate about health-care reform with a plan to further reduce the number of uninsured Americans by (1) offering coverage to people who qualify for Medicaid but live in states that rejected federal funds to broaden Medicaid eligibility after Obamacare was implemented and (2) increasing federal subsidies to individuals and families so they don't have to pay more than 8.5 percent of their income to purchase health insurance.*

But candidate Biden didn't stop there: He also endorsed a public option as an alternative to the private plans on the insurance exchanges. This version of the public option would offer lower premiums than private plans because the covered health-care services (including drugs) would be paid for at negotiated rates based on the collective purchasing power of the entire program rather than simply mirroring private insurance rates. However, this version continues to meet strong resistance because, although it would result in savings to consumers and the government, it would at the same time — and by equal measure — mean losses to the drug companies, insurance companies, and hospitals. In other words, if enacted, this would be the first major foray beyond those safe boundaries of win-win reform into the realm of zero-sum reform.

Not surprisingly, in the summer of 2020 Big Pharma countered candidate Biden's push into this contentious territory with a defensive PR campaign. A new organization, called Partnership for America's Health Care Future, funded by those same vested interests — drug companies, insurance companies, and hospitals — spent more than $9 million advertising against candidate Biden's proposal. Their website warned ominously: "Economists agree that any new government-controlled health insurance system could burden American families with unaffordable costs, in exchange for less choice and lower quality care."

As expected, the vested interests continued to defend their ongoing (albeit excessive) profitability, using their resources to convince Americans that a public option would increase health-care costs and restrict choice. The truth was just the opposite: A public option would challenge the market to function efficiently by allowing people to decide for themselves which plan best suits their needs.

But even before President Biden's inauguration, his proposal was

---

* A family of four earning $110,000 per year would save approximately $750 per month, or half its current insurance premium.

tempered to more closely resemble Medicaid Lite than a true public option. Yes, it would offer no-cost coverage to people who qualify for Medicaid but live in states that had opted not to expand their Medicaid programs. And it would be available to low-income people who could not afford to buy insurance on the exchanges. But it would not be available to most people covered by insurance at work, so it would not offer a competitive alternative for Americans who earned more than a low income. The first legislation that President Biden signed, the American Rescue Plan, included $34 billion in federal subsidies to expand health-care coverage for two years by contributing to the purchase of private insurance — a clear continuation of health reform designed *not* to challenge the vested interests.

In April 2021, PhRMA launched another defensive PR campaign — "Don't take us for granted" — with the goal of leveraging the reputational boost that followed the rollout of effective COVID-19 vaccines. PhRMA was trying to kill two birds with one stone. Defeating the Democratic plan to trim $450 billion over ten years from government payments for prescription drugs by creating a price ceiling for Medicare's highest cost-generating drugs would, at the same time, choke off funds that could have been available to pay for expansion of a true public option or lower the age of eligibility for Medicare — both of which would potentially lower the prices paid to drug companies. Steve Ubl, CEO of PhRMA, offered counterproposals with far less price relief, and added, "What we won't accept is [a] radical, hyper-partisan proposal that's going to devastate the industry." (See chapter 8 for a discussion of why the Democrats' plan to rein in at least some drug prices would have minimal negative impact on the drug companies' capacity to develop drugs that are true therapeutic advances.)

One week after PhRMA launched its "Don't take us for granted" campaign, President Biden unveiled his updated approach to health-care reform in the proposed "American Families Plan" legislation. PhRMA had achieved its goal: President Biden's proposal did not include the $450 billion cut in spending on drugs over the next ten years. And with that omission went the money that could have been used to fund an effective public option or expand Medicare. Both would have threatened not only Big Pharma's revenues but also those of their PR buddies — private insurance plans, hospitals, and, sometimes, doctors.

To the left of the public option — and rapidly fading as a realistic policy goal in the foreseeable future — is "Medicare for All." This plan would fund health care for all Americans through a tax-financed Medicare-like publicly administered program as advocated by presidential candidate Senator Bernie Sanders (I-Vermont). Many health-care progressives favor Medicare for All because it could achieve universal coverage for all Americans, cut the cost of health care by creating a consolidated purchaser with tremendous negotiating power, and save money on the wasteful administrative cost of our current disjointed hodgepodge of health-care coverage. (The administrative cost of Medicare versus private plans is 3.5 percent and 12.3 percent of plan cost, respectively.)

But the inability of Vermont governor Peter Shumlin to successfully implement a single-payer health-care system similar to Medicare for All after running and winning on that platform delivered a powerful dose of reality. Vermont's state legislature had passed the plan in 2011 with strong support from progressive health-care advocates, who saw Obamacare's abandonment of the public option as unnecessary capitulation to the private insurance industry. Disappointingly, as implementation progressed, projections showed total health-care spending would be reduced by only 1.5 percent and the increase in state taxes required to replace insurance premiums would be politically prohibitive.* Governor Shumlin withdrew the plan at the end of 2014.

A postmortem analysis of Vermont's failure to adopt a single-payer system concluded that, despite the smaller savings and

---

* A payroll tax of 14.2 percent (employers paying 10.6 percent and employees paying 3.6 percent) and additional state income tax up to 9.5 percent would have been required.

higher taxes, the plan would still have provided net savings for the 90 percent of Vermonters earning less than $150,000 per year. Ultimately, the erosion of political support caused by the projected increase in state taxes sank the plan. And therein lies an important lesson: Building a firm and realistically informed constituency for reform must be done *before* trying to legislate changes.

As American health care currently stands, the Obamacare approach of expanding coverage and appeasing the vested interests cannot be extrapolated to universal coverage because, as President Obama made clear in 2009, the cost to federal and state budgets would be too high. This leaves Medicare for All, as attractive as it is to many Americans, politically unviable until the overall cost of our health care is brought under better control.

But none of the proposals now being considered address the underlying commercial distortion of the scientific evidence that informs our health care. They do nothing to restrict the freedom the drug industry currently enjoys to manipulate research, prevent transparency, and undermine market efficiency. Pharma has done a masterful job of creating the illusion that it is the partner of physicians, nurse practitioners, and physician assistants, all of them working together to better serve patients. The truth is, Pharma is no more their partner than a car salesman is the partner of a car buyer: Sometimes their interests overlap, but their motives fundamentally differ. Many of the examples in this book were chosen to demonstrate that difference.

STEP 2: ENSURE THE RELEVANCE, ACCURACY, AND COMPLETENESS OF SCIENTIFIC EVIDENCE

If health-care reform is to succeed, the fundamental purpose of new medical knowledge must be transformed. Rather than serving primarily as a marketing tool for Big Pharma and other health-care industries, it must serve the public good, with its primary function being the cost-effective improvement of Americans' health. As shown in chapter 8, the entities that participate in the Pharma-led nexus that now produces and delivers this information currently benefit, but society as a whole does not. The following five measures would repair this failure in the integrity of commercially sponsored clinical research.

1. Rebalance What Medical Research Is About

Allowing the market to allocate most of our biomedical research funds produces a body of scientific evidence that reflects the goal of the market — to maximize financial return on investment — but is almost completely uninterested in addressing Americans' actual health needs. Now 96 percent of U.S. spending on biomedical research is about new drugs and medical devices, yet no more than 20 percent of our longevity is determined by medical care. Conversely, only 4 percent of U.S. medical research investigates the "health and well-being of individuals, communities," and the nation. Because the scientific evidence that reaches doctors and other health-care professionals is disproportionately weighted toward new drugs and devices, our health-care resources get pulled in this direction, rather than toward the overall improvement of Americans' health.

If we are to move toward a rational balance, our medical research agenda must better address the health needs of all Americans. The current mismatch is a classic instance of market failure requiring government intervention to restore and protect the interests of society as a whole. But at present our government will not even allow consideration of the cost-effectiveness of therapies in federally funded guidelines and determination of coverage, much less refocus the allocation of research funds to better address Americans' health needs.

One path forward might be to establish an independent federal health board, as suggested by Federal Reserve Board chair Ben Bernanke and supported by both Democratic (Baucus) and Republican (Grassley and Crapo) senators (discussed in chapter 9). Leaving the medical research agenda to the market has brought us data that support the greatest financial return on investment but not the greatest improvement in health. But we must be realistic about the enormous resistance that industry will mount against any attempt to diminish its control of research allocation. This applies to all the suggestions that follow and will be addressed in the final chapter.

2. Design Research to Determine Optimal Care

The fundamental purpose of clinical research must be to determine optimal care. Without this commitment, the results of clinical research, which are received by physicians and other health-care professionals as valid scientific evidence, function as little more than marketing tools. Of course, drug companies will not make this commitment on their own. Pfizer's breathtakingly honest declaration, discussed in chapter 6, shows that the purpose of data produced by its clinical trials "is to support, directly or indirectly, marketing of our product."

Because Big Pharma's primary responsibility is to serve the financial interests of its investors, companies conduct research that creates the *greatest perception of benefit* of their drugs and then sell them at the highest price possible. If, however, oversight of this market established rules of competition designed to produce the *greatest benefit to society*, the primary goal of Big Pharma's research would be to determine the most effective *and* efficient therapy. And this would require testing new products against the best available therapy.

In fact, according to the world's most widely accepted standards of research ethics, the Declaration of Helsinki Ethical Principles for Medical Research, "The benefits, risks, burdens and effectiveness of a new intervention *must be tested against those of the best proven intervention(s)*" (italics mine). (The declaration, an outgrowth of the Nuremberg trials, was designed to prevent any repetition of the human experiments conducted in Nazi concentration camps.) This unambiguous principle is critically important for two reasons. First, it protects the people who volunteer for clinical trials from receiving less than the best available care they could receive if not in a clinical trial. And second, it acknowledges the risks taken by volunteers in clinical trials by designing studies to make the greatest contribution possible to medical knowledge.

But this bedrock principle of research ethics is routinely violated. For example, as discussed in chapter 6, none of the five studies of Humira for the treatment of rheumatoid arthritis included in the current FDA-approved label compared Humira plus methotrexate to what was at the time the best proven therapy, triple therapy (methotrexate plus two other conventional drugs), which cost less than 2 percent as much. Even nineteen years after its initial FDA approval, Humira has not been tested directly against triple therapy in a randomized trial to determine if there is a difference in benefit and harms that would justify the more than fiftyfold difference in cost.

Another common failure to meet this ethical standard is not including healthy lifestyle intervention as a therapeutic option in clinical trials. The publicly funded Diabetes Prevention Program study described in chapter 4 shows how this can be done and how important the results can be. People at high risk of developing diabetes were randomly assigned to three groups — placebo, drug therapy, and lifestyle modification — and the results did not simply determine whether drug treatment reduced the risk of developing diabetes better than placebo (it did). It showed that engagement in healthy lifestyle counseling is significantly more effective than either drug therapy or placebo, which is critically important clinical knowledge, given that the same lifestyle modifications also reduce the risk of heart attack, stroke, cancer, and many other diseases. Yet studies of expensive diabetes drugs, like Trulicity, discussed in chapter 8, routinely don't bother to compare the benefit of the sponsor's drug to the benefit of lifestyle intervention.

In the spring of 2020, the COVID-19 pandemic was out of control in the United States; deaths were climbing, and six companies were racing to develop a vaccine. Yet the opportunity to determine the safest and most effective vaccine(s) was squandered. The standard approach to clinical testing was for each manufacturer to conduct its own studies, hoping to gain FDA approval by demonstrating its vaccine to be safe and significantly more protective than placebo. But government funding of these critically important studies created a unique opportunity to determine not just which vaccines were better than placebo, but which, if any, were the safest and most effective among the multiple candidates.

With government encouragement and supervision, the manufacturers could have combined their separate clinical studies into a single "master protocol." This approach — supported globally by the World Health Organization and in the United States by Dr. Anthony Fauci — called for testing all the candidate vaccines in a single trial, using a single placebo group. Entering all candidate vaccines into a single master protocol would have most efficiently determined which provided significant protection against COVID-19 infection compared to placebo; it was also the only way to determine whether any vaccines were more effective or safer than the others.*

But it was not to be. The conventional uncoordinated approach to testing had a much lower threshold of success. The risk of a vaccine being found effective compared to placebo (that is, nothing) but inferior compared to one or more of the other vaccines was avoided, allowing more winners into the marketplace.

Finally, if drug trials are to provide information about the most effective and efficient therapeutic approach, the population included in the studies must reflect the populations to which the results will be applied; the dose of the study and comparator drugs need to re-

---

* Once one or more vaccines are determined by the FDA to be effective, new trials might have to demonstrate non-inferiority to the already approved vaccines. This could require larger, more expensive trials, creating a barrier to pursuit of potentially superior candidate vaccines.

flect optimal use; and the results must be analyzed according to the protocol that was established before the trial began.

3. Require that All Authors, Peer Reviewers, and Medical Journal Editors Have Access to Complete Clinical Trial Data Prior to Publication

Real science, *especially science with both health implications and commercial value,* cannot be based on hidden data. Because the doctors' paradigm calls for acceptance of peer-reviewed results of well-designed studies published in reputable journals as scientific evidence, verification of these results *before* publication is essential. This would require that all authors, peer reviewers, and journal editors have access to the underlying data from clinical trials reported in submitted manuscripts *before* they are approved for publication. Comprehensive clinical study reports are already produced for commercially sponsored clinical trials, so making these available (with any necessary redactions to protect patient confidentiality or proprietary manufacturing information — both rare) would cause research sponsors little additional burden. The current charade of implying that peer reviewers and medical journal editors have prepublication access to the underlying data from the clinical trials reported in their pages *when they do not* must stop.*

Readers interested in the transparency of clinical trial data may be aware of the AllTrials campaign, the goal of which is to ensure that all clinical trials are registered prior to enrollment of the first patient and that a summary of study results is posted on an online registry (like clinicaltrials.gov) within one year of a study's completion. This campaign, however, risks being counterproductive by

---

* Raw individual patient-level data should be available to reviewers and editors upon request. Expertise at evaluation of data could become an academic discipline in medical or public health schools or could be provided by a government agency of health technology assessment.

creating the illusion that compliance with its requirements would adequately remedy the current lack of transparency in clinical trial reporting. This assumption fails in two ways. First, a 2020 article published in *The Lancet* found that compliance with the requirement to post study results within one year of completion "is poor, and not improving."* Second, even if studies were registered and results posted as required by law, drug companies' posting of unaudited summary results on internet registries is hardly a substitute for genuine transparency and independent analysis of clinical trial data.† Currently, efforts by the International Committee of Medical Journal Editors (discussed in chapter 8) and the AllTrials campaign risk standing in the way of data transparency by creating the false impression that progress is being made when it is not.

In May 2021, a joint statement issued by the International Coalition of Medicines Regulatory Authorities, ICMRA (including the United States, presumably in conjunction with the FDA), and the World Health Organization called for "wide access" to the clinical study reports for all new drugs and vaccines. The statement said access to these reports was necessary "to support regulators and health authorities in their decision-making; to support healthcare professionals in their treatment decisions; and to support public confidence in the vaccines and therapeutics being deployed." Lest this raise expectation of imminent change to the FDA's policy on public access to clinical study reports, an FDA spokesperson's response to an inquiry by *STAT News* showed no sense of urgency: "The FDA understands the joint statement issued by ICMRA and WHO to be strategic and aspirational and not a statement of policy."

The currently accepted lack of transparency is an abject viola-

---

* Dr. Ben Goldacre, the prime mover in the AllTrials campaign, is a coauthor on this paper.

† On its website, AllTrials.net does call for trial sponsors to make full clinical study reports publicly available, which would be a major step toward genuine transparency. The impact would, however, be minimal unless the CSRs were made available to peer reviewers prior to approving publication in a medical journal.

tion of the principle of independent verification of scientific evidence adopted more than 350 years ago by the Royal Society of London; it's a regression from Enlightenment-era science, a return to the Dark Ages, when science was subservient to external authority. The only difference: The authority of capital has replaced the authority of the church.

4. Ensure That Direct-to-Consumer Ads Inform Rather Than Mislead

Direct-to-consumer (DTC) advertising of prescription drugs is protected in the United States as a constitutional right of free speech. New Zealand is the only other country that allows DTC ads, but the two countries' widely divergent approaches to national pharmaceutical policy produce diametrically opposite results. New Zealand's Pharmaceutical Management Agency ("PHARMAC") relies on health technology assessment and national price negotiations to determine which drugs are sufficiently effective and cost efficient to merit inclusion in their national formulary. The United States, on the other hand, does not use health technology assessment to inform its national pharmaceutical policy, does not employ national (or even Medicare) price negotiations, and does not allow cost-effectiveness to be considered in federal pharmaceutical policy decisions and guidelines. Not surprisingly, among wealthy nations the United States has by far the highest per capita prescription drug costs, while New Zealand has the lowest.

The FDA monitors DTC ads (sporadically) to make sure the claims made in the ads are consistent with the information included in the FDA-approved product labels. But ensuring compliance with the *letter of the law* is a far cry from ensuring that the ads communicate a truthful and balanced understanding of the comparative benefit, risk, and cost of the advertised drug. Rather, the ads use soothing music, pleasant colors, and visuals depicting happy social activities to evoke pleasant emotions. But they fail to report such basic facts as how many people need to be treated each year for one person to benefit, how much the drug actually costs (not just how low the co-pay may be), and whether other drug or nondrug therapies are more cost-effective.

A recent ad for Trulicity, the non-insulin injectable drug to treat type 2 diabetes discussed in chapter 8, provides an excellent example. The ad makes three claims: "Most people taking [Trulicity] reached a HgbA1c under 7%," but the FDA-approved label shows that glucose control is no better than that achieved with metformin, and Trulicity costs as much as 170 times more than metformin. "Trulicity may help you lose up to 10 lbs," but the FDA-approved label shows weight loss is no greater with Trulicity than with metformin. "And [Trulicity may also help] lower your risk of cardiovascular events," but this doesn't inform viewers that the FDA-approved labels shows the benefit is so small that 99.7 percent of the people treated with Trulicity each year will not receive this benefit. None of the claims in the Trulicity ad are false, but all are misleading. They create an exaggerated appearance of superiority for the expensive new drug.

Humira, the drug most heavily advertised to consumers between 2013 and 2020, deserves special mention. Concurrent with its dominating DTC ad budget are its nearly unprecedented price increases — almost tripling since 2013, from an annual cost of $28,000 to almost $78,000 in 2021. The annual cost of Humira now exceeds the median American family's annual income, though the drug has not been proven superior for treatment of rheumatoid arthritis (its primary use) to therapy with a combination of conventional drugs that costs about one-sixtieth as much ($6,463 versus $105 per month). Nonetheless, Humira has been by far the greatest revenue-generating drug in the United States each year since 2015.

In this country, drug companies have the right to bombard us with as many advertisements as they want to pay for. But the public ought to have the right to demand that the ads contain accurate and relevant information instead of compelling but unsubstantiated emotional impressions.

5. Provide Government Oversight to Ensure a Well-Functioning Pharmaceutical Market

In his 1962 book *Capitalism and Freedom,* economist Milton Friedman argued that minimizing government intervention and market regulation would maximize personal freedom. Friedman wrote (as discussed in chapter 5) that the role of government should be limited to three major functions: "to preserve law and order, to enforce private contracts, [and] to foster competitive markets." Although considered radically conservative at the time, the idea of laissez-faire capitalism gained respect and was brought into the political mainstream by President Reagan. This marked a major turning point for the role of corporations in American society.

If the U.S. pharmaceutical industry were now to be judged by Friedman's ideals, he would look like a radical progressive. With regard to preserving law and order, pharmaceutical companies have racked up a total of over $38 billion in fines and settlements since 1991, but these punishments have amounted to no more than a very acceptable cost of doing business. For example, in 2007, when Purdue Pharma pleaded guilty to a felony for illegal marketing of its potentially deadly drug OxyContin, it was fined a mere $600 million, and the three executives involved were allowed to plead guilty to misdemeanors that did not send them to jail. Thirteen years later, after tens of thousands more U.S. deaths due to OxyContin-related overdoses, Purdue also pleaded guilty to three felonies, but this time no individuals were even charged with crimes. Clearly, the maintaining of law and order exists in appearance but not in fact.

If the government were serious about stopping drug companies' illegal marketing activities, two simple steps would go a long way. First, hold individuals who break the law personally responsible for their actions. Is extracting billions of dollars from working Americans based on false claims about drug safety or efficacy *less* of a crime than stealing a loaf of bread or passing a counterfeit $20 bill in a convenience store? A few Big Pharma executives sitting in jail, serving time for such violations, would surely have a sobering effect on drugmakers' future risk/benefit analyses related to achieving higher profits by breaking the law. And second, stop allowing big drug companies to let their subsidiaries plead guilty to felonies so that parent companies will not be "debarred" from participating in Medicare and Medicaid. Does this approach seem radical? It's consistent with the position of the free-market conservative Milton Friedman: One of government's three primary functions should be to maintain law and order.

With regard to enforcing private contracts: When drugs are prescribed and purchased, it is reasonable to expect that they will provide the benefits that the manufacturers claim they will. But manufacturers hold the real data and selectively release the information that will enhance sales. Without a formal mechanism of health technology assessment (HTA; see the next section) *with full access to clinical trial data*, doctors, patients, and insurers have no way of knowing the true value of the drugs or whether manufacturers have fulfilled their implied warranties. Government oversight through HTA would ensure that prescribers and patients receive "the benefit of the bargain." And, radical as it it may appear today, this approach would be entirely consistent with the recommendations of the most widely respected free-market economist of half a century ago.

Finally, with regard to fostering competitive markets, the consumers in Adam Smith's classical market personally evaluated the quality and price of goods to make purchases that best fulfilled their needs. But our current pharmaceutical market has no such self-regulating mechanism: The science is hidden, the published results are often overseen by the study sponsors themselves, and much of the information that reaches health-care professionals has been curated to support marketing. Moreover, total price is unregulated and usually not a factor in drug choice, and consumers' co-pays are often manipulated by middlemen (pharmaceutical benefits managers) who are susceptible to the influence of kickbacks (euphemistically called "rebates") from the manufacturers.

In his testimony of March 23, 2021, to the U.S. Senate Committee on Health, Education, Labor, and Pensions, Harvard Medical School professor Aaron Kesselheim explained that negotiating between purchasers of prescription drugs (including the government) "based on additional patient benefit, limited price increases, and stronger patent scrutiny incentivizes what matters most: the development of drugs providing important new benefits to patients and addressing unmet need." We need well-functioning markets to incentivize the development of new products that are truly beneficial (as determined by fully transparent health technology assessment) and sold at a fair price. But Big Pharma will do everything in its power to stand in the way of reforms that threaten its bottom line.

STEP 3: IMPLEMENT HEALTH TECHNOLOGY ASSESSMENT AND INDEPENDENT CLINICAL PRACTICE GUIDELINES

An estimated two-thirds of the difference in health-care spending between the United States and other wealthy countries occurs in just four areas: prescription drugs, expensive high-volume procedures, high-tech imaging like CT and MRI scans, and administrative costs. The last three categories are significant, and cost savings in those areas cannot be ignored. But pharmaceuticals contribute most to our excess spending. Even more important than the high *price* of prescription drugs here is the high *volume* of expensive drugs that are prescribed.

The uniquely high revenues generated by prescription drugs in the United States create a self-perpetuating cycle: High prices provide surplus funds, which are used in part for marketing and lobbying, which maintain the manufacturers' control of the knowledge and pricing, which increases profits to fund the next round of even more expensive new drugs, and so on. To break this cycle, healthcare professionals, payers, and the public need accurate evaluation of both the clinical and economic value of new therapies. Nearly fifty years ago, we had the means to achieve this. The U.S. Office of Technology Assessment (OTA) — the first in the world — was established in 1972 to provide "competent, unbiased information concerning the physical, biological, economic, social and political effects" of new technologies. In 1975, a health program was established within the OTA to evaluate drugs, vaccines, and other health technologies. European countries subsequently adopted our model of health technology assessment, or HTA. But in 1995, the U.S. Office of Technology Assessment was abruptly discontinued, having fallen victim to Speaker of the House Newt Gingrich's anti-regulation "Contract with America." Since then we have had no government agency to make policy recommendations based on unbiased evaluation of the comparative efficacy and price of medical innovation. In fact, we have quite the opposite: U.S. federal agencies are specifically prohibited from considering the price of drugs in coverage decisions.

The UK's government-funded National Institute for Health and Care Excellence (NICE) provides a model of HTA that might be adapted for use in the United States. NICE was originally established in 1999 to promote "clinical excellence in the health service" of England and Wales. NICE conducts appraisals of technologies (including new drugs) that include coverage recommendations, and it formulates clinical practice guidelines from the dual perspective of clinical benefit *and* cost. The UK's National Health Service is obligated to provide coverage for drugs and treatments approved by NICE.

Although fully funded by the government through the National Health Service, NICE remains at arm's length from both government and industry influence. But, before assuming the grass is perfectly green on the other side of the pond, two important caveats: First, NICE's assessments and guidelines are limited by its not having access to underlying clinical trial data. For example, NICE guidelines for the prescribing of cholesterol-lowering statin therapy to people not at high risk of cardiovascular disease rely upon the summary results of clinical trials published in medical journals rather than independent analysis of all the data from the clinical trials. And second, NICE invites patient organizations to contribute to their technology assessments. But four out of five such organizations have financial ties with manufacturers of the drugs and technologies being evaluated, and the majority are not required to report their conflicts of interest.

Notwithstanding this vulnerability to commercial influence, the role that NICE plays in appraising the value of new drugs is illustrated by its approach to the cystic fibrosis drugs manufactured by Vertex (discussed in chapter 8). In the United States, the cost of treating a person with these drugs is approximately $300,000 per year. NICE, however, recommended not covering Vertex's drugs until two conditions were met. First, it required that a lower price be negotiated; this was achieved in 2019, but the agreed-upon price remains confidential lest the rest of the world learn what the company is *really* willing to sell its drugs for. And second, NICE identified important outcomes of Vertex's clinical trials that remain unknown — like the effect on breathing capacity and quality of life after one year of therapy — and insisted that these be clarified by giving NICE ongoing access to the manufacturer's clinical trial data and the actual experience of UK citizens.

In stark contrast, federally funded coverage decisions and guidelines in the United States currently assess only "comparative effectiveness," which is based on publicly available reports of clinical trial outcomes *without regard for cost*. Moreover, the recent defunding of the National Guideline Clearinghouse, after it adopted a policy of evaluating conformity of practice guidelines with the recommendations of the National Academy of Medicine, took this country's health care a giant step back from the "arm's length" independence from industry needed to bring the best of medical science to patients.

In the United States, insurers that purchase drugs, including state Medicaid programs, are left in the compromised position of having to rely on unaudited "dossiers" prepared by manufacturers and designed to showcase the advantages of their drugs. While they appear to be informative, the purpose of these dossiers is fundamentally commercial, much like the branding and key messages discussed in chapter 7. The absence of a federally funded HTA organization in the United States has been a great boon to the biotech industry. Efforts to assess the medical and economic value of new drugs have been consigned to nongovernmental organizations that must rely largely on published results of manufacturer-sponsored clinical trials, drug company–funded cost-effectiveness analyses (as in those dossiers), and, when available, data included in FDA reports. Although inadequate, this is the best cost-effectiveness data that health-care professionals and insurers can get.

Some Americans may be leery of quasi-governmental oversight of the cost-effectiveness of their health care, so let's briefly compare our system with that of the UK. Despite the fact that we Americans spend 72 percent more of our GDP on health care, people in the UK have a 25 percent *lower* chance of dying from a cause that could have been prevented by good medical care; also, they live an average of 2.7 years longer than we do. The Commonwealth Fund ranks the UK's health-care system performance first among eleven wealthy countries, whereas our country is ranked last.* Not only are people in the UK a lot healthier than we are, but their lower health-care expenditures allow them to invest half again more of their GDP in the nonmedical determinants of health and well-being.†

---

* Rankings are based on care process, access, administrative efficiency, equity, and health outcomes.

† UK spending on social services is 20.4 percent of GDP versus 13.3 percent of GDP in the United States.

THE CHALLENGE OF ENACTING
MEANINGFUL HEALTH-CARE REFORM

Offering a wish list of reforms is easy, but overcoming the obstacles to the implementation of those reforms is not — each will be met with maximum resistance from industry (just doing its job). For its part, Congress will have a hard enough time further expanding health-care coverage by increasing subsidies to non-wealthy Americans, even with a portion of the federal investment benefiting the legislators' commercial patrons. Senator Grassley summarized the unwillingness of Congress to stand up to the vested interests on the issue of cost containment at the 2008 Health Reform Summit (discussed in the previous chapter): "We know ... that Congress does not seem to have the political guts to do anything about it now."

If Congress doesn't have the guts to stand up to Big Pharma and the other health-care industries, who does? The answer is simple but not easy. Success will require that *all stakeholders* — including doctors and other health-care providers, large and small non-health-care-related businesses, unions, government entities at all levels that pay for health care, and consumers — act in alliance to overcome the drug companies' control of the medical knowledge they now produce, manipulate, and distribute. The building of this coalition for reform is the subject of the next, and final, chapter.

11

REFORM FROM THE BOTTOM UP

The hard fact is that the way to health-care reform in the USA requires political activism of the most basic kind, something that is far beyond the comfort zone of many health professionals.

— DAVID BLUMENTHAL AND MARGARET HAMBURG,
The Lancet

From the New Deal through the 1970s, the federal government supported organizations that protected American workers, farmers, small businesses, and investors from the one-sided economic power of large corporations and financial institutions. The balance maintained by these countervailing powers reduced the need for government intervention in our economy and set the stage for our era of "blue collar aristocracy." Corporate profits rose, and the income growth of working-class households kept pace with that of the wealthiest 5 percent. The health of Americans was better than the average of the other wealthy nations, and we were not an outlier on health-care costs. Not surprisingly, during most of these years, about three-quarters of Americans trusted government "to do what was right."

But starting in the mid-1960s and accelerating since, the underpinnings of this (albeit imperfect) social contract have been whittled down, largely by three factors: the narrowing of corporations' primary mission from attending to the interests of *all* their stakeholders to maximizing profits for their shareholders,* the increasing political influence of money from corporations and the super-wealthy, and Americans' diminishing trust in government (in 2019, it was near an all-time low of 17 percent). These changes cleared the way for large corporations and very wealthy individuals to impose their will on the rest of society.

The most obvious manifestation of their greed has been the growth in inequality of income and wealth in the United States since 1980, rising above that of all other wealthy nations. A less obvious manifestation has been the unequal distribution of the fruits of medical research between the commercial sponsors (receiving large financial benefit) and the public (receiving a far less-than-commensurate health benefit). The result has been entirely predictable. Despite the expanded health-care coverage of Obamacare, our citizens' longevity has continued to lose ground in comparison to longevity in other wealthy countries, and pre-pandemic Americans' life spans actually declined in absolute terms.

As we've noted, the price of U.S. health care dramatically exceeds that of other wealthy countries, and pharmaceuticals are the largest single source of this higher cost. Of course, drug companies need to make a fair profit, but they go much further, generating huge surplus revenues. They do this in large part by widening the difference between the *claimed or implied* value and the *true* value of their new drugs. This continually growing gap doesn't just line the pockets of the drug companies — it also misleads health-care professionals by distorting the knowledge they rely on to inform patient care.

To accomplish this, Big Pharma has used two basic strategies. First, as discussed in chapter 7, they have monetized their control over the beliefs they impart about the clinical benefit of their prod-

---

* The Business Roundtable's 2019 updated statement of corporate purpose notwithstanding; see https://www.nytimes.com/2020/09/18/opinion/mil ton-friedman-essay.html.

ucts. They now set the research agenda, control the design of clinical research, own the clinical trial data (which they keep as proprietary secrets), publish the trial results in (non-transparently) peer-reviewed medical journals, and skillfully use misleading marketing and public relations campaigns. As described in chapter 6, this asymmetry of information grew as the drug companies took control of clinical trials away from academic medical centers in the 1990s. And as described in chapter 10, the banishment of our first-in-the-world government-funded health technology assessment in 1995 and the ongoing lack of HTA have removed the guardrails that, in other wealthy nations, limit the distortion of the information delivered by the commercial sponsors of medical research to the primary consumers of that so-called medical knowledge: health-care professionals.

Big Pharma's second strategy has been to raise drug prices relentlessly and unconscionably. Retail spending on prescription drugs in the United States has increased thirtyfold since 1980,* and between 2008 and 2015, the cost of the most frequently prescribed brand-name prescription drugs rose *fourteen times faster* than the Consumer Price Index. Prior to the COVID-19 pandemic, three-quarters of Americans held the opinion that the drug companies care "more about profits than they [do] about patients," and Big Pharma had earned the dubious distinction of being the most poorly regarded of American industries. This raises the question: Why haven't Pharma's scientific distortion and extortionate pricing, especially in the context of such strong negative public opinion, triggered effective policy to rebalance commercial and public interests?

The answer, in a nutshell, is that the three primary constituencies that could merge into an effective countervailing force for reform have not yet realized their potential to do so. These constit-

---

* During that period, retail sales increased from $12 billion to $358 billion (in current dollars), and this does not even include the very expensive drugs administered in hospitals and doctors' offices.

uencies include, first, health-care professionals; second, payers (non-Pharma-related businesses large and small, government, and unions); and third, health-care consumers.

It may seem at this juncture that the odds of such a coalition prevailing over the drug companies are as unlikely as David's odds of prevailing over Goliath, and that's just the way the health-care industries would like to keep it. But these constituencies *can* pool their influence to create a formidable force — capable of controlling costs, providing universal coverage, and improving the health of Americans to at least the level enjoyed by citizens of other wealthy countries. Understanding how this can happen requires a deeper look at each of these constituencies and its potential influence on reform.

The first constituency comprises the doctors, nurses, and other health-care professionals, whose primary responsibility is to bring the best of medical science to their patients' care. Overwhelmingly respected by patients, these health-care workers are the natural leaders of this coalition.

Among these potential activists, doctors play a unique role as both the gatekeepers and the primary consumers of scientific evidence about new drugs. But without adequate oversight of the integrity of clinical trials, fully transparent peer review before publication, clinical practice guidelines that are based on full access to clinical trial data and free of commercial influence, and independent, federally funded health technology assessment, doctors lack the basic tools they need to fulfill their responsibility as learned intermediaries.

Health-care professionals' most urgent — and most personally challenging — task in reform must, therefore, be to recognize the extent to which their own beliefs about optimal practice have been and continue to be manipulated by the drug industry's distortion of scientific evidence delivered through the channels they have been trained to trust. Once this is understood, the need for the reforms presented in chapter 10 will become painfully apparent. At that point, health-care professionals will *insist* on reforms that apply the fundamental rules of good science to the knowledge that guides their practice.

An example of such an effort is the Choosing Wisely campaign, an initiative of the American Board of Internal Medicine Foundation. Its purpose is to minimize low-value care (for example, ordering spine X-rays and other imaging to evaluate routine short-term back pain). Hopefully, the Choosing Wisely campaign will soon demand access to full clinical trial data, so it can broaden its efforts to also address low-value prescribing, as well as help health-care professionals learn how to interpret primary data and thereby become more aware of the discrepancies in the information that they have heretofore been accepting at face value.

Health-care professionals' leadership role comes, however, with another great responsibility: making their congressional representatives, the media, and the public aware of the need for these reforms. And most important, they need to engage their patients in partnerships based on what will truly make them healthier (including lifestyle changes) rather than simply ordering more tests and prescribing the newest medications. They also need to consider the cost of that care in determining optimal therapy, and they must reject the Pharma-serving meme that insensitivity to the price of the drugs they prescribe is in their patients' best interest.

Our doctors, nurses, and other health-care workers have emerged as the heroes of the COVID-19 pandemic. They are now even more empowered to lead the movement to improve Americans' health care and health. Expecting health-care professionals to include activism among their already enormous responsibilities may seem like too much additional burden. However, many will find that a constructive and coordinated response to the burnout and "moral injury" they routinely suffer as participants in our dysfunctional health-care system is far more fulfilling than passive (if disgruntled) acceptance of the status quo.

The second constituency is the payers who foot the bill for our

wasteful health care: the employers and unions, as well as government entities at all levels, that provide health-care benefits. Their
primary goal is to receive fair value for their health-care spending;
however, given the current excessive cost and overuse of new and
unnecessarily expensive prescription drugs (as well as high-volume
lucrative procedures and imaging studies), that goal remains unre-
alized. So, like health-care professionals, the purchasers of health
care must insist on independent, federally funded HTA to ensure,
on the one hand, that they *do* receive fair value for their health-
care dollars and, on the other, that the people they cover receive the
most effective health care.

Payers must also insist that the oversight of the quality of medical care that they're purchasing — provided by organizations such
as the National Committee for Quality Assurance (NCQA) and the
Joint Commission on Accreditation of Healthcare Organizations
(JCAHO) — be *completely* free of commercial sponsorship.* Furthermore, purchasers must use their considerable power to pressure Congress to control drug prices, offer a public option to all
employees, and support noncommercially influenced research to
study the most effective ways to improve population health.

Last of the three constituencies — but very much not least — are
health-care consumers, meaning all of us, and we want to be as
healthy as possible. Like doctors, we consumers must understand
that most of what we think we know about brand-name drugs has
come directly or indirectly from the manufacturers and has been
carefully curated to increase demand for their products. Heavily
advertised drugs are being pushed because of their *financial* value
to the manufacturers, not their *health* value to us.

More fundamentally, Big Pharma's intense marketing seduces us
into assuming that we can achieve optimal health simply by taking

---

* Chapter 4 discusses NCQA-endorsed and industry-favorable standards of
care for people with type 2 diabetes, and chapter 7 points out JCAHO's financial relationship with Purdue Pharma.

a pill, obscuring the fact that our health is mostly determined not by medical care but by how we live our lives. Remember, observational data from the Nurses' Health Study show that more than 80 percent of heart disease and diabetes could be prevented by adherence to simple, commonsense healthy habits. We have more control over our health than we have been led to believe, and also more responsibility for it.

But — and this is a big "but" — socioeconomic disparities can make adoption of healthy lifestyle habits a difficult, and even insurmountable, challenge for many Americans. If we are really serious about improving the health of *all* Americans, then we consumers must advocate for a medical research agenda that reflects *that* goal instead of the current one, which focuses almost exclusively on expensive new drugs and devices. We will not be able to achieve this objective without bringing our excessive spending on medical care and our underspending on the social determinants of health into a better balance.

Some of us will have a harder time than others adopting healthy lifestyle habits, but all of us can vote. We as consumers, along with the other two constituencies, must make our voices heard by voting for representatives who will support these changes in Congress and our statehouses. But voting is not enough. Most legislators receive campaign contributions from Pharma and are the target of its now record-breaking lobbying efforts, so consumers need to make sure that their representatives understand how *strongly* they feel about prescription-drug prices and the need to regulate what has become a rogue industry.

The path toward meaningful and lasting reform begins with this tripartite coalition, but it will also require government oversight — actively supported by this coalition — to defend the public interest. The formation of an independent federal health board (modeled after the Federal Reserve Board) will be essential to ensuring the relevance, integrity, transparency, and accuracy of clinical research; to creating and overseeing a federal health technology assessment agency; and to negotiating and regulating the price of drugs and other medical technologies.

Health care has become yet another test of our democracy. Although success is in no way preordained, we, the citizens of that democracy, *can* rise to meet these immense challenges.

AFTERWORD

As I finish this manuscript, the need for health-care reform in the United States is becoming even more acute, but the headwinds are growing ever stronger. The challenges described in this book — the excesses of the pharmaceutical industry; the complicity of academic researchers and institutions, journals, and professional societies; the inability of doctors, other health-care professionals, and the public to understand the extent to which they are being manipulated by commercial interests; and the failure of government to provide adequate oversight to protect the public's interest — are by no means relics of a dysfunctional past.

The COVID-19 pandemic presented a stress test for the health-care systems of all nations, but particularly that of the United States. Between 2018 and the end of 2020, Americans' longevity declined by 1.9 years — a drop of over eight times more than the average decline in sixteen other wealthy countries. This decline highlights the disparities in our system: Longevity for Black and Hispanic Americans declined two and a half to three times more than it did for whites. By the end of 2020, longevity for Black men had fallen to 67.7 years, all the way back to where it had been in 1998. Although the vaccines that were rushed to market have saved countless lives and prevented enormous suffering, their rollout has been marred by racial disparities: As of the end of June 2021, white Americans were 40 percent more likely to have received at least one COVID-19 vaccination than Black Americans were.

These grim statistics highlight an uncomfortable truth: The claim that health care in the United States is superior to that of other wealthy countries is just not true. And even worse, our health-care priorities are accelerating in the wrong direction.

In June 2021 the FDA made what appears to be its worst approval decision ever, giving the green light to Aduhelm, a new drug to treat Alzheimer's disease. The facts about its lack of efficacy and safety are not in dispute: The manufacturer stopped its two major clinical trials of Aduhelm, EMERGE and ENGAGE, in 2019 because interim analyses of the trial data showed "futility," meaning the studies were statistically very unlikely to demonstrate clinical benefit. Months later, post hoc reanalysis found patients who had been treated with high-dose Aduhelm in one of the two abandoned studies had experienced *statistically* significant improvement on the Clinical Dementia Rating scale. But these patients had improved by only 0.39 points on the 18-point scale, a mere fraction of the 1- to 2-point improvement that is the minimum considered to be *clinically* significant. Furthermore, in reporting the combined results from the EMERGE and ENGAGE studies, Biogen's own safety data showed that brain swelling or bleeding was visible in the MRI scans of 41 percent of the patients treated with Aduhelm, compared to only 9 percent of the patients treated with placebo. In other words, one out of three *additional patients* treated with Aduhelm at the dose approved by the FDA developed either brain swelling or bleeding. Similarly, almost one out of ten additional patients treated with Aduhelm developed symptoms (including headache, dizziness, visual disturbances, and nausea and vomiting) that correlated with these MRI abnormalities.*

The FDA's statistical reviewer concluded, "In summary, the totality of the data does not seem to support the efficacy" of Aduhelm. And when the FDA's Advisory Committee was asked whether the

---

* MRI abnormalities and symptoms were present in 10 percent of patients treated with Aduhelm, compared to 0.6 percent of those treated with placebo. The absolute risk difference associated with the combined high-dose Aduhelm groups from both studies versus placebo was 9.4 percent (95 percent confidence interval 7.6, 11.2), calculated on openepi.com.

results of the single so-called positive study provided "primary evidence of the effectiveness of [Aduhelm]," their vote was nearly unanimous: Ten voted no, one voted uncertain, and zero voted yes. Shortly after this, the FDA announced its approval of Aduhelm, overruling the recommendation of its Advisory Committee; three committee members resigned.

So what was the FDA thinking? To approve Aduhelm, it had to ignore its own statistical reviewer, Biogen's own safety data, and the Advisory Committee's virtually unanimous vote. Nonetheless, Janet Woodcock, the acting FDA commissioner, explained that Biogen's studies had shown that Aduhelm reduces the size of amyloid plaques in the brain, plaques that correlate with Alzheimer's disease, and she was "fairly confident" that improvement in this surrogate end point would lead to clinical benefit. Just one problem: All twenty-seven previous studies exploring the clinical effect of drugs shown to reduce amyloid deposits had either failed to show clinical benefit or had shown toxicity. And, as it turns out, the FDA was fully aware of this, as revealed in its 2018 draft guidance for industry about drug development for Alzheimer's disease. This document states that there was "no sufficiently reliable evidence" that the response to treatment of any biomarker (such as amyloid deposits) "would be reasonably likely to predict clinical benefit."

When the FDA's decision to ignore the Advisory Committee's recommendation produced a public outcry, Dr. Patrizia Cavazzoni, the head of the FDA's Center for Drug Effectiveness and Research, which oversees the approval of new drugs, offered a proposal to stave off any such disagreeable reactions in the future. Dr. Cavazzoni — who had joined the FDA in 2018, after sixteen years spent working for major pharmaceutical companies — suggested that the way to "reduce the emotional" aspects of debate and minimize surprises for manufacturers at the Advisory Committee meetings was to loosen the "overly restrictive" rules against financial ties between drug companies and committee members.

In my opinion, increasing the number of committee members getting paid by industry may be a fantastic idea if you are a biotech executive or investor, but not so good if you are an ordinary working American hoping for better health.

If the insanity of treating hundreds of thousands, if not millions, of Alzheimer's patients with a drug for which there is no solid evidence of benefit but for which there is solid evidence of *harm* is not stupefying enough, consider this: The FDA's decision is projected to create a monumental financial burden for Medicare because Alzheimer's disease is unfortunately common among people sixty-five years of age and older, and because Biogen has priced its drug at an appalling $56,000 per year. Using the lower end of Biogen's projections concerning the number of people expected to use this drug, which is one million of the six million Americans with Alzheimer's disease, Aduhelm alone could increase Medicare's spending on prescription drugs administered in hospitals or doctors' offices by 150 percent (adding $57 billion annually* to Medicare Part B's current spending, which is $37 billion).

In a rational health-care system, the FDA's approval of a drug like Aduhelm — not shown to provide meaningful clinical benefit, causing side effects for a significant number of patients, and rejected by the FDA Advisory Committee — would set off alarms. Instead, we are on track to spend tens of billions of dollars annually on this latest iteration of biotech snake oil.

This situation reveals yet another impediment to reform. A survey of Americans aware of the FDA's approval of Aduhelm found that 60 percent believed — despite the overwhelming lack of evidence — that "this drug will be effective." Such irrational optimism brings to mind the immortal words of Walt Kelly's cartoon character Pogo: "We have met the enemy and he is us." It's another example of our "phish"-ability — our tendency to want to accept a marketing claim at face value without understanding our reasons for wanting to believe it, and Big Pharma's readiness to exploit this vulnerability again and again.

---

* This does not include the hefty cost of the MRI and PET scans required for screening and follow-up monitoring.

The wasting of Americans' money and the squandering of our health *can* be stopped, but only if all of us — from the wealthiest drug company CEO to the heads of the agencies charged with overseeing our health care, to our trusted health-care professionals, to the hundreds of millions of the "rest of us" — recognize an urgent fact: Our health and the quality of our lives are being sacrificed to the insatiable corporate and investor lust for health-care profits. After all, not even the greedy perpetrators are immune to its harmful effects. Just like the physician who thought he was doing Stacey Palmer a favor by giving her samples of Vioxx, the "best" possible care provided by a well-meaning doctor to a drug company CEO will be compromised by information that has been distorted for commercial purposes, potentially putting that CEO's health at risk.

We are all sitting, side by side, in the same boat. Our difficult task is to turn that boat around and set it on a better course: toward the goal of optimal health at reasonable cost for every American.

ACKNOWLEDGMENTS

I want to remember Bruce Spitz, who introduced me to the pleasures and pitfalls of health policy research in 1981. We remained close friends and colleagues until he passed away in 2006. I also want to remember Wolf Kahn, Emily Mason, Jim McConkey, and Charlotte Maurer, all of whom generously shared their wisdom.

My agent and friend Kris Dahl has been a supportive guide for almost twenty years.

Alex Littlefield, my editor at Houghton Mifflin Harcourt, remained unflinchingly committed to this project from the beginning. His response to the first draft of this manuscript was enormously encouraging — he expressed "admiration and gratitude" for my work. This was followed by an eleven-page single-spaced editorial memo that explained its deficiencies. Several days later, when the defensive voices inside my head quieted enough to let me consider what Alex was saying, I realized he was right. The steps that followed were not easy, but they were productive as the final structure emerged. Olivia Bartz joined Alex on this project when I submitted a revised manuscript, and I am grateful to both of them for feedback and patience as the book took its final form. I also want to acknowledge Fariza Hawke for her assistance with the final manuscript, and especially Deborah Jacobs and Laura Brady, who were meticulous with the final proofreading and production details.

Special thanks to Patsy Shepherd, who read, edited, and discussed each word of each version of this work over nearly six years. She has become a trusted friend and a highly respected colleague.

And big thanks to Tracy Roe, who copyedited my book. Besides copyediting, Tracy has another job — urgent care physician. She treated me and my manuscript with no less care than a committed physician treats a patient. I am also most grateful to Susanna Brougham, who copyedited a second round of revisions, offering important guidance and expert polish. Other editorial assistance for which I am grateful was provided by Sonni Efron, Will Murphy, and David Newman.

The investigative journalist Jeanne Lenzer deserves special mention; her work showed me, more than two decades ago when I was a full-time family physician, that integrity, courage, and persistence can stand up to even the most powerful forces.

Many thanks to George Akerlof, who kindly (and patiently) assisted me in integrating the principles of behavioral economics into health-care professionals' vulnerability to pharmaceutical marketing.

Special gratitude goes to the hardworking and thoughtful behind-the-scenes curators of several blogs and listservs that provide real-time discussion of unfolding issues related to corporate distortion of medical knowledge and our health care: David Healy (davidhealy.org), Roy Poses (hcrenewal.blogspot.com), and my colleagues at the Biojest listserv. In addition, thanks to Mark Wilson, Juan Gervas, and Mohammad Zakaria Pezeshki, whose efforts to send me relevant and timely developments in the medical literature have been most helpful.

Much appreciation to my medical and legal colleagues who shared insights, source documents, and critical feedback: Dharma Cortés, Clint Fisher, Laura Fitzpatrick, Colleen Fuller, Isabel Goldman, Norton Hadler, Kristen Johnson, Robert M. Kaplan, Tom Sobol, Mark Wilson (again), and Jim Wright.

And thanks to the following friends (listed in alphabetical order) for contributing ideas, feedback, and support: Judy Abrams and Sherman Kelly, Jean Abramson and Deb Barnard, Joel Altschul, Carol Austin and Tom Jarvis, Judy Azulay, Michael Baker, Jeff Berkowitz, Steven Bloomstein, Rob Boone, Janice and Jim Boyko, Hannah Brandes and Herb Brown, Margaret Crone, Rick Dana, David Elan, Ron Fox, Hank Greenspan, Cathy Kahn and Ralph Laumbach, Michael Koleda, Leo Liu, Elliot Lobel, Darby Mumford, Julie Nagazina and Richard Einhorn, Michael and Paula Nathanson, Connie Richardson (in memory), Michael Schaaf, Anne and Paul Spirn, Crawford Taisey, Bruce Vladek, Roger Warner, Joan Yospin, and all my friends from Bustins Island, Maine.

Special thanks to my ninety-eight-year-old mother-in-law, Ruth Kahn (author of *My Father's Dragon*), who kept me fortified with her authorial empathy and oatmeal cookies.

NOTES

INTRODUCTION

page

xi   *inferior health and health care:* Bradley Sawyer and Daniel McDermott, "How Do Mortality Rates in the U.S. Compare to Other Countries?," Health System Tracker, February 14, 2019, https://www.healthsys temtracker.org/chart-collection/mortality-rates-u-s-compare-coun tries/#item-leading-causes-death-mortality-rates-higher-u-s-compara ble-countries.

   provided by Obamacare: Sawyer and McDermott, "How Do Mortality Rates."

xii   *of just 10.7 percent:* Rabah Kamal et al., "Health Spending and the Econ- omy," Health System Tracker, https://www.healthsystemtracker.org/ indicator/spending/health-expenditure-gdp/.

   to fill the gap: Kimford J. Meador, "Decline of Clinical Research in Aca- demic Medical Centers," *Neurology* 85 (2015): 1171–76.

xiii   *unavailable medical benefits:* "Drug Industry: Profits, Research and De- velopment Spending, and Merger and Acquisition Deals," Report to Con- gressional Requesters, U.S. Government Accountability Office, Novem- ber 2017, https://www.gao.gov/assets/690/688472.pdf.

   global pharmaceutical profits: Dana Goldman and Darius Lakdawalla, "The Global Burden of Medical Innovation," USC Schaeffer, Leonard D. Schaeffer Center for Health Policy and Economics, January 2018, 2–3, https://healthpolicy.usc.edu/wp-content/uploads/2018/01/01.2018_ Global20Burden20of20Medical20Innovation.pdf.

xiv   *survey began, in 2001:* Justin McCarthy, "Big Pharma Sinks to the Bot- tom of U.S. Industry Rankings," *Gallup,* September 3, 2019, https://news.

gallup.com/poll/266060/big-pharma-sinks-bottom-industry-rankings.
aspx.

not panned out financially: Lucy Hooker and Daniele Palumbo, "Covid Vaccines: Will Drug Companies Make Bumper Profits?," BBC Business, https://www.bbc.com/news/business-55170756.

($36 versus $7.25): Chris Bickerton, "Europe Failed Miserably with Vaccines. Of Course It Did," *New York Times,* May 17, 2021.

of the European Union: Bickerton, "Europe Failed Miserably."

(one dose of vaccine): Monika Pronczuk and Léontine Gallois, "Pfizer and Moderna Raised Their Vaccine Prices in Their Latest E.U. Contracts," *New York Times,* August 2, 2021.

xv   *top ten among U.S.:* Keith Speights, "Here's 1 Surprising Ranking Where Moderna Handily Beats Pfizer," *Motley Fool,* May 18, 2021.

begin large clinical trials: Selam Gebrekidan and Matt Apuzzo, "Rich Countries Signed Away a Chance to Vaccinate the World," *New York Times,* March 21, 2021.

"Big Pharma started on": "Pfizer Vaccine Relies on U.S. Government–Developed Spike Protein Technology," *Public Citizen,* November 10, 2020.

xvi   *"Unfortunately some in Congress":* PhRMA, FINAL_PhRMA_Print_American-Success-Story1.pdf.

"used unusual contracts": Gebrekidan and Apuzzo, "Rich Countries Signed Away," updated May 7, 2021.

as low as $3 per shot: "$25 Billion to Vaccinate the World," *Public Citizen,* May 24, 2021.

$33.5 billion in 2021: Zachary Brennan, "When a Life-Saving Vaccine Became a Cash Cow: Pfizer Now Projects $33.5B in Covid-19 Vaccine Sales in 2021 Alone," *EndpointsNews,* July 28, 2021.

xvii *"in the high 20 percent":* Rebecca Robbins and Peter S. Goodman, "Pfizer Reaps Hundreds of Millions in Profits from Covid Vaccine," *New York Times,* updated May 5, 2021.

"a very, very marginal": Eben Shapiro, "Pfizer CEO Albert Bourla Raises Expectations That the Pharmaceutical Giant Can Deliver a COVID-19 Vaccine by Fall," *Time,* July 12, 2020.

"a significant opportunity": Lee Fang, "Drugmakers Promise Investors They'll Soon Hike COVID-19 Vaccine Prices," *The Intercept,* March 18, 2021.

the next five years: Allana Akhtar, "Pfizer Could Sell Nearly $100 Billion Worth of COVID-19 Vaccines in the Next Five Years, Morgan Stanley Estimates," *Business Insider,* May 10, 2021.

60 to 80 percent: Arlene Weintraub, "Pfizer CEO Says It's 'Radical' to

Suggest Pharma Should Forgo Profits on COVID-19 Vaccine: Report,"
FiercePharma, July 30, 2020.

by over $50 billion: People's Vaccine — Vaccine Billionaires press release
— 20 May 2021.pdf, https://app.box.com/s/c487wmiyquh9q1glpbatzf
5sukls7ph2.

xviii *in low-income countries:* Jesus Jiménez and Andrés R. Martínez, "The
Rush to Vaccinate the World Stalls as Funds and Doses Fall Short," *New
York Times,* June 6, 2021.

"a scandalous inequity": Nicholas Kristof, "Vaccinate the World! The
Best Investment Ever," *New York Times,* May 26, 2021.

and vaccine makers: Emily Baumgaertner, "Vaccine Companies and the
U.S. Government Snubbed WHO Initiative to Scale Up Global Manufacturing," *Los Angeles Times,* April 30, 2021.

"That's an easy answer": Interview with Dr. Anthony Fauci, *Partners in Health,* January 12, 2021, https://twitter.com/PIH/status/
1349155249911713792.

"At this point in time": Ed Silverman, "Pharma Leaders Shoot Down
WHO Voluntary Pool for Patent Rights on COVID-19 Products," *STAT
News,* May 28, 2020.

"We feel a special obligation": "Statement by Moderna on Intellectual
Property Matters During the COVID-19 Pandemic," October 8, 2020.

"taken no steps to share": Rachel Silverman, "Waiving Vaccine Patents
Won't Help Inoculate Poorer Nations," *Washington Post,* March 15, 2021.

xix *impending negative consequences:* Kristalina Georgieva, Tedros Adha-
nom Ghebreyesus, David Malpass, and Ngozi Okonjo-Iweala, "Here's
Our Plan to Increase Vaccine Access and End the Pandemic Faster,"
Washington Post, May 31, 2021.

in developing countries: Chelsea Clinton and Achal Prabhala, "The Vaccine Donations Aren't Enough: The Pandemic Won't Be Over Until It's
Under Control Around the World," *Atlantic,* June 20, 2021.

"deadly variants to emerge": World Health Organization, "A Global Pandemic Requires a World Effort to End It — None of Us Will Be Safe Until
Everyone Is Safe," September 30, 2020.

"at least 60 percent": Georgieva et al., "Here's Our Plan."

vaccines for developing countries: Ruchir Agarwal and Gita Gopinath, "A
Proposal to End the COVID-19 Pandemic," *IMF,* May 19, 2021.

"the highest return on": Kristof, "Vaccinate the World!"

increased tax revenues: Agarwal and Gopinath, "A Proposal to End."

nonwealthy countries would: Robbins and Goodman, "Pfizer Reaps Hundreds of Millions."

xx   *"one of the great public"*: Robbins and Goodman, "Pfizer Reaps Hundreds of Millions."

xxi   *the U.S. death rate:* Sawyer and McDermott, "How Do Mortality Rates."

xxii *an op-ed I wrote:* John Abramson, "Information Is the Best Medicine," *New York Times,* September 18, 2004.

      the biggest drug recall: See https://www.seegerweiss.com/drug-injury/vioxx/.

xxiv *"with the intent to defraud"*: "Justice Department Announces Largest Health Care Fraud Settlement in Its History: Pfizer to Pay $2.3 Billion for Fraudulent Marketing," Department of Justice, Office of Public Affairs, September 2, 2009, https://www.justice.gov/opa/pr/justice-department-announces-largest-health-care-fraud-settlement-its-history.

1. VIOXX: AN AMERICAN TRAGEDY

1   *"Don't believe anything"*: Quoted in Richard Knox, "Merck Tries to Move Beyond Vioxx Debacle," *Morning Edition,* National Public Radio, November 12, 2007, https://www.npr.org/templates/story/story.php?storyId=16211947?storyId=16211947.

3   *doubled the risk:* Robert S. Bresalier et al., "Cardiovascular Events Associated with Rofecoxib in a Colorectal Adenoma Chemoprevention Trial," *New England Journal of Medicine* 352 (2005): 1092–1102.

      "unexpected increase in": Gina Kolata, "A Widely Used Arthritis Drug Is Withdrawn," *New York Times,* October 1, 2004, http://www.nytimes.com/2004/10/01/business/a-widely-used-arthritis-drug-is-withdrawn.html.

      "clearly there" and "a shame": Edward Scolnick, internal Merck e-mail, March 9, 2000, in Vioxx Litigation Documents collection, UCSF Industry Documents Library, https://www.industrydocumentslibrary.ucsf.edu/drug/docs/jsww0217.

4   *"significantly fewer serious"*: Garret A. FitzGerald and Carlo Patrano, "The Coxibs, Selective Inhibitors of Cyclooxygenase-2," *New England Journal of Medicine* 345 (2001): 433–42.

      "The difference in major": FitzGerald and Patrano, "The Coxibs."

5   *disregard of their own rules:* Jeffrey M. Drazen and Gregory D. Curfman, "Financial Associations of Authors," *New England Journal of Medicine* 346 (2002):1901–2.

6   *no better pain relief:* Shari Targum, FDA medical officer, Memorandum: "Review of Cardiovascular Safety Database," February 1, 2001, 10, https://web.archive.org/web/20180127041414/https://www.fda.gov/ohrms/dockets/ac/01/briefing/3677b2_06_cardio.pdf.

in the Washington Post: S. Okie, "Missing Data on Celebrex: Full Study Altered Picture of Drug," *Washington Post,* August 5, 2001.

7   *Celecoxib Long-Term Arthritis:* Fred E. Silverstein et al., "Gastrointestinal Toxicity with Celecoxib vs. Nonsteroidal Anti-inflammatory Drugs for Osteoarthritis and Rheumatoid Arthritis," *Journal of the American Medical Association* 284 (2000): 1247–55.

    older anti-inflammatory drugs: FDA Arthritis Advisory Committee, "Briefing Information. Celebrex (celecoxib)," February 7, 2001, https://web.archive.org/web/20010806212823/http://www.fda.gov/ohrms/dockets/ac/01/briefing/3677b1.htm.

    "My concern is that": David Armstrong, "How the *New England Journal* Missed Warning Signs on Vioxx," *Wall Street Journal,* May 15, 2006.

8   *"the potentially serious":* Thomas W. Abrams to Raymond V. Gilmartin, president and CEO of Merck, warning letter, September 17, 2001, https://web.archive.org/web/20050825 104239/http://www.fda.gov/cder/warn/2001/9456.pdf.

    into the journal's coffers: Armstrong, "How the *New England Journal.*"

    the previous five years: Duff Wilson, "Merck to Pay $950 Million over Vioxx," *New York Times,* November 22, 2011.

9   *forms of cyclooxygenase:* Byron Cryer and Andre Dubois, "The Advent of Highly Selective Inhibitors of Cyclooxygenase — a Review," *Prostaglandins and Other Lipid Mediators* 56 (1998): 341–61.

10 *"E-Mails Suggest Merck Knew":* Anna Wilde Mathews and Barbara Martinez, "E-Mails Suggest Merck Knew Vioxx's Dangers at Early Stage," *Wall Street Journal,* November 1, 2004.

    people taking low-dose aspirin: Patients taking aspirin, including those taking low-dose aspirin because of a history of cardiovascular disease, were excluded from the VIGOR study. Even so, after the study was completed, 4 percent of study participants were found to have met the FDA's criteria for, but were not taking, low-dose aspirin.

    five years on the U.S. market: David J. Graham et al., "Risk of Acute Myocardial Infarction and Sudden Cardiac Death in Patients Treated with Cyclo-Oxygenase 2 Selective and Non-Selective Non-Steroidal Anti-Inflammatory Drugs: Nested Case-Control Study," *Lancet* 365 (2005): 475–81.

15 *increased risk of heart attack:* Gregory D. Curfman, Stephen Morrissey, and Jeffrey M. Drazen, "Expression of Concern," *New England Journal of Medicine* 353 (2005): 2813–14.

    Seattle pharmacist and Dr. Drazen: Armstrong, "How the *New England Journal.*"

    "The story is playing out": Armstrong, "How the *New England Journal.*"

17 *any company in history:* Knox, "Merck Tries to Move Beyond."
    "in the forefront of": Gardner Harris, "F.D.A. Official Admits 'Lapses' on Vioxx," *New York Times,* March 2, 2005.

18 *almost twenty-seven thousand:* Meredith Wadman, "Merck Settles Vioxx Lawsuits for $4.85 Billion," Nature.com, November 13, 2007, https://www.nature.com/articles/450324b.
    "false statements about": "U.S. Pharmaceutical Company Merck Sharp & Dohme to Pay Nearly One Billion Dollars over Promotion of Vioxx," Office of Public Affairs, Department of Justice, November 22, 2011, https://www.justice.gov/opa/pr/us-pharmaceutical-company-merck-sharp-dohme-pay-nearly-one-billion-dollars-over-promotion.
    "The central allegation": David Gelles, "Merck C.E.O. Ken Frazier on Death Row Cases and the Corporate Soul," *New York Times,* March 9, 2018, https://www.nytimes.com/2018/03/09/business/merck-ceo-ken-frazier-on-death-row-cases-and-the-corporate-soul.html.

2. NEURONTIN: FRAUD AND RACKETEERING

21 *that year was Neurontin:* David C. Radley, Stan N. Finkelstein, and Randall S. Stafford, "Off-Label Prescribing Among Office-Based Physicians," *Archives of Internal Medicine* 166 (2006): 1021–26.
    "little or no scientific support": Randall S. Stafford, "Regulating Off-Label Drug Use — Rethinking the Role of the FDA," *New England Journal of Medicine* 358 (2008): 1427–29.

22 *"the 'snake oil'":* Jim Edwards, "Lesson from Pfizer: Don't Describe Your Product as 'Snake Oil' in Internal Email," *CBS News,* March 26, 2010, https://www.cbsnews.com/news/lesson-from-pfizer-dont-describe-your-product-as-snake-oil-in-internal-email/.
    reached $2.1 billion: See https://www.drugs.com/top200_2003.html.

23 *Pfizer's off-label marketing:* John Abramson, "Expert Report of John Abramson, MD," United States District Court, February 2010, 45, https://www.communitycatalyst.org/pal-docs/neurontin_exh_N.pdf.
    86 percent of worldwide: Abramson, "Expert Report," 44.

25 *not responding adequately:* Abramson, "Expert Report," 120.
    those treated with placebo: Abramson, "Expert Report," 102–4.
    an independent study: Atul C. Pande et al., "Gabapentin in Bipolar Disorder: A Placebo-Controlled Trial of Adjunctive Therapy," *Bipolar Disorders* 2 (2000): 249–55.
    no better than placebo: Abramson, "Expert Report," 56–57.
    increased fiftyfold: Abramson, "Expert Report," 129.

26   *psychiatry teleconferences:* Abramson, "Expert Report," 102.

more than eleven thousand: Abramson, "Expert Report," 105–6.

Psychiatric and Mental Health Congress: Abramson, "Expert Report," 103.

the Young study: Abramson, "Expert Report," 116–17.

27   *worse than nothing:* Abramson, "Expert Report," 119–20.

"Our job is not to go": Abramson, "Expert Report," 117–18.

"The doctors are on their own": Abramson, "Expert Report," 136.

"fraudulent marketing": Thomas Sullivan, "Supreme Court Rejection of Pfizer's Request for RICO Off Label Review: Could Open Floodgate of Cases," *Policy and Medicine,* May 6, 2018, https://www.policymed.com/2014/01/supreme-court-rejection-of-pfizers-request-for-rico-off-label-review-could-open-floodgate-of-cases.html.

the Gorson study: Abramson, "Expert Report," 149.

29   *violating its own study:* Abramson, "Expert Report," 142–45.

"May I interrupt you": Abramson, "Expert Report," 142.

30   *than a sugar pill:* Abramson, "Expert Report," 17. The study actually considered four types of pain, or four end points. Only one of these end points showed that Neurontin was significantly more effective than placebo.

"have you essentially": Abramson, "Expert Report," 145.

31   *reduced daily pain scores:* Miroslav Backonja et al., "Gabapentin for the Symptomatic Treatment of Painful Neuropathy in Patients with Diabetes Mellitus: A Randomized Controlled Trial," *Journal of the American Medical Association* 280 (1998): 1831–36.

"educate both consumers and": Abramson, "Expert Report," 118.

15 percent in the placebo group: Abramson, "Expert Report," 23. Dizziness occurred in 23.8 percent of people in the Neurontin group (compared to 4.9 percent in the placebo group), sleepiness in 22.6 percent of people in the Neurontin group (versus 6.2 percent in the placebo group), and headaches in 10.7 percent of the Neurontin group (compared to 3.7 percent in the placebo group).

As I explained to the jury: Abramson, "Expert Report," 21, slide 32.

32   *"did not account for":* Backonja et al., "Gabapentin for the Symptomatic Treatment."

33   *two biostatistics journals:* Farid Jamshidian, Alan E. Hubbard, and Nicholas P. Jewell, "Accounting for Perception, Placebo, and Unmasking Effects in Estimating Treatment Effects in Randomised Clinical Trials," *Statistical Methods in Medical Research,* e-published September 8, 2011; Alan Hubbard, Farid Jamshidian, and Nicholas Jewell, "Adjusting for

Perception and Unmasking Effects in Longitudinal Clinical Trials," *International Journal of Biostatistics* 8, no. 2 (2012): 1–20.

"there was no statistically": Abramson, "Expert Report," 66.

"Although I would love": Abramson, "Expert Report," 32, slide 42.

34   *"should not be pushed"*: Abramson, "Expert Report," 32, slide 46.

as an independent article: Abramson, "Expert Report," 37.

to file the application: Abramson, "Expert Report," 35.

"substantial evidence against": Abramson, "Expert Report," 35–37.

35   relief at a lower dose: Abramson, "Expert Report," 43–44.

"Gabapentin doses up to": Abramson, "Expert Report," 92.

"At doses of 1800 to 3600": Miroslav Backonja and Robert I. Glanzman, "Gabapentin Dosing for Neuropathic Pain: Evidence from Randomized, Placebo-Controlled Clinical Trials," *Clinical Therapeutics* 25, no. 1 (2003): 81–104.

36   *"Is it legal to promote"*: Abramson, "Expert Report," 52.

"Well, that is really important": Abramson, "Expert Report," 55.

37   *tripled the penalty*: "Jury Finds Pfizer Committed Fraud and Racketeering Through Off-Label Neurontin Promotions; Awards $142 Million," *Health Policy Hub*, April 18, 2010, https://www.communitycatalyst.org/blog/jury-finds-pfizer-committed-fraud-and-racketeering-through-off-label-neurontin-promotions-awards-142-million-pal#.W28kZyhKhPY.

"nationwide effort to unlawfully": Judge Saris, "Amended Findings of Fact and Conclusions of Law," United States District Court: District of Massachusetts, August 31, 2011, 1, https://www.gpo.gov/fdsys/pkg/USCOURTS-mad-1_04-cv-10739/pdf/USCOURTS-mad-1_04-cv-10739-7.pdf.

38   *"clinicians who prescribe"*: Christopher W. Goodman and Allan S. Brett, "A Clinical Overview of Off-Label Use of Gabapentinoid Drugs," *JAMA Internal Medicine* 179 (2019): 695–701.

sixth most frequently prescribed: "Top 10 Prescription Medications in the U.S. (August 2020)," GoodRx, https://www.goodrx.com/drug-guide.

3. THE TRUTH ABOUT STATINS

39   *most frequently prescribed*: "Selected Prescription Drug Classes Used in the Past 30 Days, by Sex and Age: United States, Selected Years 1988–1994 Through 2011–2014," National Center for Health Statistics, 2016, https://www.cdc.gov/nchs/data/hus/2016/080.pdf.

42   *attempts at lifestyle modification*: Expert Panel on Detection, Evaluation, and Treatment of High Blood Cholesterol in Adults, "Executive

Summary of the Third Report of the National Cholesterol Education Program (NCEP) Expert Panel on Detection, Evaluation, and Treatment of High Blood Cholesterol in Adults (Adult Treatment Panel III)," *Journal of the American Medical Association* 285 (2001): 2486–97, doi:10.1001/jama.285.19.2486.

43 *"favorable effects" in women:* Expert Panel, "Executive Summary."
"clinical trials of LDL lowering": Expert Panel on Detection, Evaluation, and Treatment of High Blood Cholesterol in Adults, "Third Report of the National Cholesterol Education Program (NCEP) Expert Panel on Detection, Evaluation, and Treatment of High Blood Cholesterol in Adults (Adult Treatment Panel III)," NCEP, National Heart, Lung, and Blood Institute, National Institutes of Health, September 2002, table 8.2-1.
not been enough women: Barbara Alving to Merrill Goozner, National Heart, Lung, and Blood Institute, U.S. Department of Health and Human Services, October 22, 2004, https://web.archive.org/web/20041101125157/https://www.nhlbi.nih.gov/guidelines/cholesterol/response.pdf.
not scientifically valid: "Protocol for a Prospective Collaborative Overview of All Current and Planned Randomized Trials of Cholesterol Treatment Regimens, Cholesterol Treatment Trialists' (CTT) Collaboration," *American Journal of Cardiology* 75 (1995): 1130–34. This article specifically identifies women as a subgroup of interest in determining the efficacy of statins.
had been 130 mg/dL: Scott M. Grundy et al., "Implications of Recent Clinical Trials for the National Cholesterol Education Program Adult Treatment Panel III Guidelines," *Circulation* 110 (2004): 227–39.

44 *no history of heart disease:* Peter S. Sever et al., "Prevention of Coronary and Stroke Events with Atorvastatin in Hypertensive Patients Who Have Average or Lower-than-Average Cholesterol Concentrations, in the Anglo-Scandinavian Cardiac Outcomes Trial – Lipid Lowering Arm (ASCOT-LLA): A Multicenter Randomised Controlled Trial," *Lancet* 361 (2003): 1149–58.
"the absence of specific": Alving to Goozner, October 22, 2004.
And there still isn't: Samia Mora et al., "Statins for the Primary Prevention of Cardiovascular Events in Women with Elevated High-Sensitivity C-Reactive Protein or Dyslipidemia: Results from the Justification for the Use of Statins in Prevention: An Intervention Trial Evaluating Rosuvastatin (JUPITER) and Meta-Analysis of Women from Primary Prevention Trials," *Circulation* 121 (2010): 1069–77.

45 *on the NHLBI website:* "ATP III Update 2004: Implications of Recent

Clinical Trials for the ATP III Guidelines Updates and Disclosures," National Heart, Lung, and Blood Institute, National Institutes of Health, U.S. Department of Health and Human Services, 2004, https://web.ar chive.org/web/20150107221016/ https://www.nhlbi.nih.gov/health-pro/ guidelines/current/cholesterol-guidelines/update-2004/disclosures.

"Are Lipid-Lowering Guidelines": John Abramson and James M. Wright, "Are Lipid-Lowering Guidelines Evidence-Based?," *Lancet* 369 (2007): 168–69.

overall risk of death: J. R. Downs et al., "Primary Prevention of Acute Coronary Events with Lovastatin in Men and Women with Average Cholesterol Levels: Results of AFCAPS/TexCAPS. Air Force/Texas Coronary Atherosclerosis Prevention Study," *Journal of the American Medical Association* 279 (1998): 1615–22.

46   *included almost a quarter:* Michael J. Pencina et al., "Application of New Cholesterol Guidelines to a Population-Based Sample," *New England Journal of Medicine* 370 (2014): 1422–31, fig. 2.

"Don't Give More Patients": John D. Abramson and Rita F. Redberg, "Don't Give More Patients Statins," *New York Times,* November 13, 2013, https://www.nytimes.com/2013/11/14/opinion/dont-give-more-pa tients-statins.html.

"We relied extensively": Jennifer Robinson, vice-chair of the 2013 cholesterol guidelines panel, Cholesterol Treatment Trialists' (CTT) Collaboration, January 7, 2016, https://www.youtube.com/watch?v=QNVtBjH caIk.

47   *into meta-analyses:* See https://www.cttcollaboration.org/.

"unequivocal evidence about": "Protocol for a Prospective Collaborative," *American Journal of Cardiology.*

48   *between ages forty and seventy-five:* Pencina et al., "Application of New Cholesterol Guidelines."

asked to lead the panel: Jeanne Lenzer, "Majority of Panelists on Controversial New Cholesterol Guideline Have Current or Recent Ties to Drug Manufacturers," *British Medical Journal* 347 (2013): f6989.

for consulting and research: "Jennifer G. Robinson," OpenPaymentsData. CMS.gov, Centers for Medicare and Medicaid Services, https://openpay mentsdata.cms.gov/physician/35650/summary.

"that chairs of guideline": Bernard Lo and Marilyn J. Field, eds., *Conflict of Interest in Medical Research, Education, and Practice* (Washington, DC: National Academies Press, 2009), 14.

vice-chairs were added: Clinical Practice Guidelines We Can Trust (Washington, DC: National Academies Press, 2011), 7.

49   *rely on industry support:* "American Heart Association Financial Information," American Heart Association, https://www.heart.org/en/about-us/aha-financial-information, and "Principles for Relationships with Industry," American College of Cardiology, https://www.acc.org/about-acc/industry-relations/principles-for-relationships-with-industry.

     CTT meta-analysis of 2012: CTT Collaborators, "The Effects of Lowering LDL Cholesterol with Statin Therapy in People at Low Risk of Vascular Disease: Meta-Analysis of Individual Data from 27 Randomised Trials," *Lancet* 380 (2012): 581–90.

51   *20 percent ten-year risk:* CTT Collaborators, "The Effects."

     "with statin therapy in people": CTT Collaborators, "The Effects."

52   *very little or no net benefit:* John D. Abramson et al., "Should People at Low Risk of Cardiovascular Disease Take a Statin?," *British Medical Journal* 347 (2013): f6123.

53   *"the panel was unanimous":* Iona Heath et al., "Report of the Independent Panel Considering the Retraction of Two Articles in the *BMJ*," *British Medical Journal* 349 (2014): g5176.

     history of cardiovascular: Ingrid Torjessen, "GPs Should Consider Offering Statins to All Patients Aged over 75, Researchers Say," *British Medical Journal* 364 (2019): l522, doi: 10.1136/bmj.l522.

     "annually, 1000 people": John Abramson et al., "Questioning Statin Therapy for Older Patients," *Lancet* 395 (June 2020): 1831–32, https://www.thelancet.com/pdfs/journals/lancet/PIIS0140-6736(19)33156-3.pdf.

     they did not dispute: Jordan Fulcher et al., "Authors' Reply," *Lancet* 395 (June 2020): 1832–33.

     from 8.8 to 34.1 percent: Gregory Curfman, "Risks of Statin Therapy in Older Adults," *JAMA Internal Medicine* 177, no. 7 (2017): 966.

54   *she gained weight:* Takehiro Sugiyama et al., "Is There Gluttony in the Time of Statins? Different Time Trends of Caloric and Fat Intake Between Statin-Users and Non-Users Among US Adults," *JAMA Internal Medicine* 174, no. 7 (2014): 1038–45.

     just three habits: "Unhealthy Diets and Physical Inactivity," NMH Fact Sheet, World Health Organization, https://www.who.int/nmh/publications/fact_sheet_diet_en.pdf.

55   *"How could the statin":* Paul D. Thompson, "What to Believe and Do About Statin-Associated Adverse Effects," *Journal of the American Medical Association* 316 (2016): 1969–70.

     experienced muscle symptoms: Beth A. Parker et al., "The Effect of Statins on Skeletal Muscle Function," *Circulation* 127, no. 1 (2013): 96–103.

people taking statins: Erik S. Stroes et al., "Statin-Associated Muscle Symptoms: Impact on Statin Therapy — European Atherosclerosis Society Consensus Panel Statement on Assessment, Aetiology, and Management," *European Heart Journal* 36, no. 17 (May 2015): 1012–22, https://doi.org/10.1093/eurheartj/ehv043.

renal failure, cataracts: Julia Hippisley-Cox and Carol Coupland, "Unintended Effects of Statins in Men and Women in England and Wales: Population Based Cohort Study Using the Qresearch Database," *British Medical Journal* 340 (2010): c2197.

neuropathy, sexual dysfunction: Beatrice A. Golomb and Marcella A. Evans, "Statin Adverse Effects: A Review of the Literature and Evidence for a Mitochondrial Mechanism," *American Journal of Cardiovascular Drugs* 8, no. 6 (2008): 373–418.

exertional fatigue: Beatrice A. Golomb et al., "Effects of Statins on Energy and Fatigue with Exertion: Results from a Randomized Controlled Trial," *Archives of Internal Medicine* 172, no. 15 (2012): 1180–82.

psychiatric symptoms: Michael Tatley and Ruth Savage, "Psychiatric Adverse Reactions with Statins, Fibrates, and Ezetimibe: Implications for the Use of Lipid-Lowering Agents," *Drug Safety* 30, no. 3 (2007): 195–201.

56   *"The statins trialists":* Emma Parish, Theodora Bloom, and Fiona Godlee, "Statins for People at Low Risk," *British Medical Journal* 351 (2015): h3908.

4. INSULIN, INC.

58   *"Nearly a century":* Jeremy A. Greene and Kevin R. Riggs, "Why Is There No Generic Insulin? Historical Origins of a Modern Problem," *New England Journal of Medicine* 372 (2015): 1171–75.

diabetes, meaning "siphon": Jacob Roberts, Science History Institute, December 8, 2015, https://www.sciencehistory.org/distillations/magazine/sickening-sweet.

diagnosed with diabetes: Centers for Disease Control and Prevention, "National Diabetes Statistics Report 2020: Estimates of Diabetes and Its Burden in the United States," U.S. Department of Health and Human Services, https://www.cdc.gov/diabetes/pdfs/data/statistics/national-diabetes-statistics-report.pdf.

treated with insulin: "Fast Facts, Data, and Statistics About Diabetes," American Diabetes Association, https://professional.diabetes.org/sites/professional.diabetes.org/files/media/fast_facts_9-2018.pdf.

60 *on the floor, dead:* Author's personal communication with Nicole Holt-Smith, February 24, 2019.

61 *improve his clinical condition:* Howard Markel, "How a Boy Became the First to Beat Back Diabetes," *PBS NewsHour*, January 11, 2013, https://www.pbs.org/newshour/health/how-a-dying-boy-became-the-first-to-beat-diabetes.

after the first day of therapy: F. G. Banting et al., "Pancreatic Extracts in the Treatment of Diabetes Mellitus," *Canadian Medical Association Journal* 12, no. 3 (March 1922): 141–46.

succumbing to pneumonia: Markel, "How a Boy."

frowned on profiting: Greene and Riggs, "Why Is There No Generic."

from pig pancreases: Louis Rosenfeld, "Insulin: Discovery and Controversy," *Clinical Chemistry* 48 (2002): 2270–88.

62 *genes of* E. coli *bacteria:* David V. Goeddel et al., "Expression in *Escherichia coli* of Chemically Synthesized Genes for Human Insulin," *Proceedings of the National Academies of Science USA* 76, no. 1 (1979): 106–10.

the same chemical structure: See https://www.gene.com/stories/biotech-basics.

had prima facie appeal: Colleen Fuller and Audrey Farley, *Diabetes, Inc.: The Clinical, Commercial, and Cultural History of a Disease* (in press).

"outstandingly pure": Greene and Riggs, "Why Is There No Generic."

63 *blood-sugar control:* Greene and Riggs, "Why Is There No Generic."

would decrease: Greene and Riggs, "Why Is There No Generic."

animal insulin once used: Kasla J. Lipska et al., "Use and Out-of-Pocket Costs of Insulin for Type 2 Diabetes Mellitus from 2000 Through 2010," *Journal of the American Medical Association* 311, no. 22 (2014): 2331–33.

64 *"clinically relevant differences":* B. Richter and G. Neises, "'Human' Insulin Versus Animal Insulin in People with Diabetes Mellitus," *Cochrane Database of Systematic Reviews* (2003), https://doi.org/10.1002/14651858.CD003816.

rather than human insulin: A. Siebenhofer et al., "Short Acting Insulin Analogues Versus Regular Human Insulin in Patients with Diabetes Mellitus," *Cochrane Database of Systematic Reviews* (2006), https://doi.org/10.1002/14651858.CD003287; Moshe Vardi and Assaf Nini, "Intermediate Acting Versus Long Acting Insulin for Type 1 Diabetes Mellitus," *Cochrane Database of Systematic Reviews* (2008), https://doi.org/10.1002/14651858.CD006297.

"almost identically effective": K. Horvath et al., "Long-Acting Insulin Analogues Versus NPH Insulin (Human Isophane Insulin) for Type 2 Dia-

betes Mellitus," *Cochrane Database of Systematic Reviews* (2007), https://
doi.org/10.1002/14651858CD005613; Siebenhofer et al., "Short Acting Insulin."

"a cautious response": Siebenhofer et al., "Short Acting Insulin."

65 *were for insulin analogs:* Lipska et al., "Use and Out-of-Pocket Costs."

the global insulin market: David Beran, Irl B. Hirsch, and John S. Yud-
kin, "Why Are We Failing to Address the Issue of Access to Insulin? A
National and Global Perspective," *Diabetes Care* 41, no. 5 (2018): 1125–31,
https://care.diabetesjournals.org/content/41/6/1125.

$21 per vial: John Russell, "Lilly Insulin Prices Come Under the Micro-
scope," *Indianapolis Business Journal,* August 25, 2017, https://www.ibj.
com/articles/65163-lilly-insulin-prices-come-under-microscope.

more than 18 percent: https://web.archive.org/web/20180707032834/
https://www.cms.gov/Newsroom/MediaReleaseDatabase/Press-re
leases/2018-Press-releases-items/2018-05-15.html?DLPage=1&DLEn
tries=10&DLSort=0&DLSortDir=descending.

$330 per vial: See https://web.archive.org/web/20170710042951/https:
//www.goodrx.com/humalog?form=vial&dosage=10ml-of-100-
units-ml&quantity=&days_supply=&label_override=Humalog.

almost certainly be alive: Alaric Dearment, "Lilly's Lower-Cost Insulin Humalog Still Too Expensive, Senator Says," *MedCity News,* March
5, 2019; Sarah Varney, "Watch: High Cost of Insulin Sends Americans
to Canada to Stock Up," *Kaiser Health News,* https://khn.org/news/insu
lin-high-cost-americans-caravan-to-canada-for-cheaper-supply/.

an average of $5,224: Xinyang Hua, Natalie Carvalho, and Michelle
Tew, "Expenditures and Prices of Antihyperglycemic Medications in the
United States: 2002–2013," *Journal of the American Medical Association*
315, no. 13 (2016): 1400–1402.

just $468 per year: Kasla J. Lipska, Irl B. Hirsch, and Matthew C. Riddle,
"Human Insulin for Type 2 Diabetes: An Effective, Less-Expensive Option," *Journal of the American Medical Association* 318, no. 1 (2017): 23–24.

66 *"reported cost-related insulin":* Darby Herkert et al., "Cost-Related Insulin Underuse Among Patients with Diabetes," *JAMA Internal Medicine*
179, no. 1 (2019): 112–14.

American Association of Clinical Endocrinologists: Yehuda Handelsman
et al., "American Association of Clinical Endocrinologists and American College of Endocrinology — Clinical Practice Guidelines for Developing a Diabetes Mellitus Comprehensive Care Plan — 2015 — Executive
Summary," *Endocrine Practice* 21, no. 4 (April 2015): 413–37, https://doi.
org/10.4158/EP15672.GL.

American Diabetes Association: J. L. Chiang et al., "Type 1 Diabetes Through the Life Span: A Position Statement of the American Diabetes Association," *Diabetes Care* 37 (2014): 2034–54, https://doi.org/10.2337/dc14-1140.

type 2 diabetes mellitus: https://professional.diabetes.org/sites/profes sional.diabetes.org/files/media/fast_facts_8-2017_pro_3.pdf.

67   *obesity, which raises the risk:* Theresia M. Schnurr et al., "Obesity, Unfavourable Lifestyle, and Genetic Risk of Type 2 Diabetes: A Case-Cohort Study," *Diabetologia* 63 (2020): 1324–32, https://doi.org/10.1007/s00125-020-05140-5.

two- to threefold: Serban Dinca-Panaitescu et al., "Diabetes Prevalence and Income: Results of the Canadian Community Health Survey," *Health Policy* 99 (2011): 116–23.

eventually require insulin: "Fast Facts: Data and Statistics About Diabetes," American Diabetes Association, https://professional.diabe tes.org/sites/professional.diabetes.org/files/media/fast_facts_8-2017_pro_3.pdf.

type 1 (1.5 million): CDC, "Statistics Report."

best place for drug companies: "Diabetes Prevalence," Health at a Glance 2017, https://doi.org/10.1787/health_glance-2017-15-en.

Countries that allowed: Sarah Elizabeth Holden et al., "Evaluation of the Incremental Cost to the National Health Service of Prescribing Analogue Insulin," *British Medical Journal Open* 1, no. 2 (2011): e000258, http://bmjopen.bmj.com/content/bmjopen/1/2/e000258.full.pdf.

68   *defined as "poor control":* Barbara B. Fleming et al., "The Diabetes Quality Improvement Project: Moving Science into Health Policy to Gain an Edge on the Diabetes Epidemic," *Diabetes Care* 24, no. 10 (2001): 1815–20.

69   *makers of those insulin analogs:* David Aron and Leonard Pogach, "Transparency Standards for Diabetes Performance Measures," *Journal of the American Medical Association* 301, no. 2 (2009): 210–12.

so-called quality standards: Aron and Pogach, "Transparency Standards."

up to $80 per patient: Andrew J. Ahmann, "Guidelines and Performance Measures for Diabetes," *American Journal of Managed Care* 13, suppl. 2 (2007): S41–S46, https://www.ajmc.com/journals/supple ment/2007/2007-04-vol13-n2suppl/apr07-2476ps41-s46.

70   *Assurance (NCQA):* Aron and Pogach, "Transparency Standards."

"associated with fewer episodes": Dawn E. DeWitt and Irl B. Hirsch, "Outpatient Insulin Therapy in Type 1 and Type 2 Diabetes Mellitus," *Journal of the American Medical Association* 289 (2003): 2254–64.

"near-normal [HbA1c] levels": Irl B. Hirsch, "Insulin Analogs," *New England Journal of Medicine* 352 (2005): 174–83.

less than 7 percent: Aron and Pogach, "Transparency Standards."

71   *less stringent standard*: Aron and Pogach, "Transparency Standards."

"It appears that a bandwagon": Aron and Pogach, "Transparency Standards."

severe hypoglycemic event: Action to Control Cardiovascular Risk in Diabetes Study Group, "Effects of Intensive Glucose Lowering in Type 2 Diabetes," *New England Journal of Medicine* 358 (2008): 2545–59.

"at least 60% of patients": 2009 Diabetes Recognition Program, "2009 Diabetes Recognition Program Requirements," National Committee for Quality Assurance, effective August 1, 2009, rev. April 1, 2011, https://rec ognitionportal.ncqa.org/documents/DRP%20Requirements.pdf.

72   *hypoglycemia and even death*: Amir Qaseem et al., "Hemoglobin A1c Targets for Glycemic Control with Pharmacologic Therapy for Nonpregnant Adults with Type 2 Diabetes Mellitus: A Guidance Statement Update from the American College of Physicians," *Annals of Internal Medicine* 168, no. 8 (2018): 569–76.

for blood-sugar control: René Rodriguez-Gutierrez et al., "Benefits and Harms of Intensive Glycemic Control in Patients with Type 2 Diabetes," *British Medical Journal* 367 (2019): l5887, http://dx.doi.org/10.1136/bmj. l5887.

recombinant human insulin: Alan J. Garber et al., "Consensus Statement by the American Association of Clinical Endocrinologists and American College of Endocrinology on the Comprehensive Type 2 Diabetes Management Algorithm — 2016 Executive Summary," *Endocrine Practice* 22, no. 1 (2016): 84–113, https://read.qxmd.com/read/26731084/ consensus-statement-by-the-american-association-of-clinical-endo crinologists-and-american-college-of-endocrinology-on-the-compre hensive-type-2-diabetes-management-algorithm-2016-executive-sum mary.

"for those who find": Alan J. Garber et al., "Consensus Statement by the American Association of Clinical Endocrinologists and American College of Endocrinology on the Comprehensive Type 2 Diabetes Management Algorithm — 2018 Executive Summary," *Endocrine Practice* 24, no. 1 (2018): 91–120, https://doi.org/10.4158/CS-2017-0153.

unexpected expense of $400: Consumer and Community Research Section, Division of Consumer and Community Affairs, "Report on the Economic Well-Being of U.S. Households in 2018," Board of Governors of the Federal Reserve System, May 2019, https://www.federalreserve.gov/pub

lications/files/2018-report-economic-well-being-us-households-201905.
pdf.

analogs over recombinant: American Diabetes Association, "Pharmacologic Approaches to Glycemic Treatment: Standards of Medical Care in Diabetes — 2018," *Diabetes Care* 41, suppl. 1 (2018): S73–S85, https://doi.org/10.2337/dc18-S008.

"control of fasting glucose": American Diabetes Association, "Pharmacologic Approaches to Glycemic Treatment: Standards of Medical Care in Diabetes — 2019," *Diabetes Care* 42, suppl. 1 (2019): S90–S102, https://doi.org/10.2337/dc19-S009.

73   *drugs, tests, or other products:* AACE, "About AACE: Corporate AACE Partnership (CAP)," American Association of Clinical Endocrinology, https://pro.aace.com/about/corporate-aace-partnership-cap.

drug or device manufacturers: American Diabetes Association, "Disclosures: Standards of Medical Care in Diabetes — 2019," *Diabetes Care* 42, suppl. 1 (2019): S184–S186, https://doi.org/10.2337/dc19-Sdis01.

74   *"probably" provide an advantage:* "Safety, Effectiveness, and Cost Effectiveness of Long Acting Versus Intermediate Acting Insulin for Patients with Type 1 Diabetes: Systematic Review and Network Meta-Analysis," *British Medical Journal* 349 (2014): g5459.

"despite the advantages": American Diabetes Association, "Pharmacologic Approaches to Glycemic Treatment: Standards of Medical Care in Diabetes — 2020," *Diabetes Care* 43, suppl. 1 (2020): S98–S110.

75   *to 26.9 million:* CDC, "Statistics Report."

among the wealthy nations: "Diabetes Prevalence," Health at a Glance.

went up only slightly: Cheryl D. Fryar, Margaret D. Carroll, and Cynthia L. Ogden, "Prevalence of Overweight, Obesity, and Extreme Obesity Among Adults Aged 20 and Over: United States, 1960–1962 Through 2013–2014," National Center for Health Statistics, July 18, 2016, https://www.cdc.gov/nchs/data/hestat/obesity_adult_13_14/obesity_adult_13_14.htm.

to 39.8 percent: Fryar, Carroll, and Ogden, "Prevalence."

average of the other countries': "Obesity Update, 2017," OECD.org, https://www.oecd.org/els/health-systems/Obesity-Update-2017.pdf.

will be obese: Zachary J. Ward et al., "Projected U.S. State-Level Prevalence of Adult Obesity and Severe Obesity," *New England Journal of Medicine* 381 (2019): 2440–50.

76   *treated with metformin:* Diabetes Prevention Program Research Group, "Reduction in the Incidence of Type 2 Diabetes with Lifestyle Intervention or Metformin," *New England Journal of Medicine* 346, no. 6 (2002): 393–403.

lower-leg amputations: See https://www.cdc.gov/diabetes/pdfs/data/statistics/national-diabetes-statistics-report.pdf.

cost for each year: William H. Herman, "The Cost-Effectiveness of Lifestyle Modification or Metformin in Preventing Type 2 Diabetes in Adults with Impaired Glucose Tolerance," *Annals of Internal Medicine* 142, no. 5 (2005): 323–32.

77   *an enormous bargain:* Peter J. Neumann, Joshua T. Cohen, and Milton C. Weinstein, "Updating Cost-Effectiveness — the Curious Resilience of the $50,000-per-QALY Threshold," *New England Journal of Medicine* 371 (2014): 796–97.

three healthy lifestyle habits: Frank B. Hu et al., "Diet, Lifestyle, and the Risk of Type 2 Diabetes Mellitus in Women," *New England Journal of Medicine* 345 (2001): 790–97.

rising pharmaceutical expenditures: Joseph L. Dieleman et al., "Factors Associated with Increases in US Health Care Spending, 1996–2013," *Journal of the American Medical Association* 318, no. 17 (2017): 1668–78.

increased spending on insulin: Hua, Carvalho, and Tew, "Expenditures."

$300 per person per year: Ronald T. Ackerman, "Working with the YMCA to Implement the Diabetes Prevention Program," *American Journal of Preventive Medicine* 44, no. 4 (2013): S352–S356.

78   *"promising early results":* Elizabeth K. Ely et al., "A National Effort to Prevent Type 2 Diabetes: Participant-Level Evaluation of CDC's National Diabetes Prevention Program," *Diabetes Care* 40, no. 10 (2017): 1331–41, https://care.diabetesjournals.org/content/40/10/1331.

79   *$133 per year:* Dzintars Gotham, Melissa J. Barber, and Andrew Hill, "Production Costs and Potential Prices for Biosimilars of Human Insulin and Insulin Analogues," *British Medical Journal Global Health* 3 (2018): e000850.

5. AS AMERICAN SOCIETY GOES

83   *"American health care":* Angus Deaton, "It's Not Just Unfair: Inequality Is a Threat to Our Governance," *New York Times Book Review,* March 20, 2017.

84   *outstanding student loans:* Michael Corkery and Stacy Cowley, "Household Debt Makes a Comeback in the U.S.," *New York Times DealBook,* May 17, 2017, https://www.nytimes.com/2017/05/17/business/dealbook/household-debt-united-states.html?_r=0.

than we were in 1980: "Health Status," OECD, stats.oecd.org, http://stats.oecd.org/Index.aspx?DataSetCode=HEALTH_STAT#.

"the money spent has": David M. Cutler, Allison B. Rosen, and Sandeep Vijan, "The Value of Medical Spending in the United States, 1960–2000," *New England Journal of Medicine* 355 (2006): 920–27.

waste, fraud, or abuse: Jonathan Gruber, "Health Care Reform," speech given at Holy Cross College, broadcast on C-SPAN, March 11, 2010.

85   *"the vast majority of"*: Austin Frakt, "Blame Technology, Not Longer Life Spans, for Health Spending Increases," *New York Times*, January 23, 2017.

ten comparable countries: Bradley Sawyer and Cynthia Cox, "How Does Health Spending in the U.S. Compare to Other Countries?," Health System Tracker, December 7, 2018, https://www.healthsystemtracker.org/chart-collection/health-spending-u-s-compare-countries/#item-average-wealthy-countries-spend-half-much-per-person-health-u-s-spends.

$1.5 trillion in excess: "Total Health Spending," Health System Tracker, https://www.healthsystemtracker.org/indicator/spending/health-expenditure-gdp/.

86   *in the other countries*: Sawyer and McDermott, "How Do Mortality Rates."

actually decreased: "U.S. Life Expectancy 1950–2021," Macrotrends, https://www.macrotrends.net/countries/USA/united-states/life-expectancy.

87   *Syria, Yemen, and Venezuela:* See https://apps.who.int/gho/data/node.main.HALE?lang=en.

"Even non-Hispanic white": Steven H. Woolf and Laudan Y. Aron, "The U.S. Health Disadvantage Relative to Other High-Income Countries: Findings from a National Research Council/Institute of Medicine Report," *Journal of the American Medical Association* 309, no. 8 (2013): 771–72.

40 percent as much: James Banks et al., "Disease and Disadvantage in the United States and in England," *Journal of the American Medical Association* 295, no. 17 (2006): 2037–45.

the Commonwealth Fund: Eric C. Schneider et al., "Mirror, Mirror 2017: International Comparison Reflects Flaws and Opportunities for Better U.S. Health Care," Commonwealth Fund, July 2017, https://interactives.commonwealthfund.org/2017/july/mirror-mirror/.

88   *"result of the development"*: Cutler, Rosen, and Vijan, "Value of Medical Spending."

89   *from 1960 to 2015:* Jiaquan Xu et al., "Deaths: Final Data for 2016," *National Vital Statistics Reports* 57, no. 5 (July 26, 2018), https://www.cdc.gov/nchs/data/nvsr/nvsr67/nvsr67_05.pdf.

among OECD countries: "Health at a Glance 2017: OECD Indicators," OECD iLibrary, oecd-ilibrary.org, November 10, 2017, https://read.oecd-ilibrary.org/social-issues-migration-health/health-at-a-glance-2017_health_glance-2017-en#page61.

less than five and a half: T. J. Matthews and Marian F. MacDorman, "Infant Mortality Statistics from the 2004 Period Linked Birth/Infant Death Data Set," *National Vital Statistics Reports* 55, no. 14 (June 13, 2007): 16, https://www.cdc.gov/nchs/data/nvsr/nvsr55/nvsr55_14.pdf.

to teenage mothers: Matthews and MacDorman, "Infant Mortality."

high-risk, low-birth-weight: Lindsay A. Thompson, David C. Goodman, and George A. Little, "Is More Neonatal Intensive Care Always Better? Insights from a Cross-National Comparison of Reproductive Care," *Pediatrics* 109, no. 6 (2002): 1036–43.

90   *"the effectiveness of the current":* Thompson, Goodman, and Little, "Is More Neonatal Intensive Care."

average of those other: Ashish P. Thakrar et al., "Child Mortality in the US and 19 OECD Comparator Nations: A 50-Year Time-Trend Analysis," *Health Affairs* 37, no. 1 (2018): 140–49, doi: 10.1377/hlthaff.2017.0767.

91   *than we had since 2000:* "Mortality from Circulatory Diseases," Health at a Glance 2019: OECD Indicators, OECD iLibrary, oecd-ilibrary.org, https://read.oecd-ilibrary.org/social-issues-migration-health/health-at-a-glance-2019_4dd50c09-en#page77.

no better in Pennsylvania: David Cutler, "Why Does Health Care Cost So Much in America? Ask Harvard's David Cutler," interview by Paul Solman, *PBS NewsHour,* November 19, 2013, https://www.pbs.org/newshour/nation/why-does-health-care-cost-so-much-in-america-ask-harvards-david-cutler.

92   *minimum-wage laws:* Thomas Kochan, "Election Rage Shows Why America Needs a New Social Contract to Ensure the Economy Works for All," *Conversation,* November 16, 2016, https://theconversation.com/election-rage-shows-why-america-needs-a-new-social-contract-to-ensure-the-economy-works-for-all-68296.

Depression-era home: Steven Pressman, "GOP Tax Plan Doubles Down on Policies That Are Crushing the Middle Class," *Conversation,* December 20, 2017.

investment in infrastructure: John Cassidy, "Forces of Divergence: Is Surging Inequality Endemic to Capitalism?," *New Yorker,* March 24, 2014, https://www.newyorker.com/magazine/2014/03/31/forces-of-divergence.

in addition to shareholders: Ralph Gomory and Richard Sylla, "The American Corporation," *Daedalus* 142, no. 2 (Spring 2013): 102–18.

"perhaps the major peacetime": Robert B. Reich, *Saving Capitalism: For the Many, Not the Few* (New York: Knopf, 2016).

the top 5 percent: "Percentage Change in Family Income, 1947–1975 and 1975–2012," *Chartbook of Social Inequality,* https://www.russellsage.org/sites/all/files/chartbook/Income%20and%20Earnings.pdf.

"Corporations have a responsibility": Gomory and Sylla, "The American Corporation."

93 *below the rate of inflation:* P. R. Vagelos, "Are Prescription Drug Prices High?," *Science* 252 (1991): 1080–84.

six and ten times faster: Immaculada Hernandez et al., "The Contribution of New Product Entry Versus Existing Product Inflation in the Rising Costs of Drugs," *Health Affairs* 38, no. 1 (2019): 76–83, doi: 10.1377/hlthaff.2018.05147; Tim McMahon, "Historical Consumer Price Index (CPI-U) Data," InflationData.com, November 12, 2020, https://inflationdata.com/Inflation/Consumer_Price_Index/HistoricalCPI.aspx?reloaded=true.

the World Health Organization: World Health Organization, "Ivermectin," African Programme for Onchocerciasis Control, http://www.who.int/apoc/cdti/ivermectin/en/.

"blue collar aristocracy": Anne Case and Angus Deaton, "Mortality and Morbidity in the 21st Century," *Brookings Papers on Economic Activity* (March 23, 2017): 397–476.

spending only slightly more: Sawyer and Cox, "How Does Health Spending."

94 *under President Carter:* Andrew Downer Crain, "Ford, Carter, and Deregulation in the 1970s," *Journal on Telecommunications and High Technology Law* 5 (2007): 413–47.

guru behind this wisdom: Lanny Ebenstein, *Milton Friedman: A Biography* (New York: St. Martin's, 2007).

capitalism with political freedom: Peter S. Goodman, "A Fresh Look at the Apostle of Free Markets," *New York Times,* April 13, 2008.

"to preserve law and order": Milton Friedman, *Capitalism and Freedom,* fortieth anniversary edition (Chicago: University of Chicago Press, 2002), 2.

greatest prosperity: Scott Lanman and Steve Matthews, "Greenspan Concedes to 'Flaw' in His Market Ideology," Bloomberg.com, October 23, 2008, https://www.bloomberg.com/news/articles/2008-10-23/green

span-concedes-to-flaw-in-his-market-ideology; Friedman, *Capitalism and Freedom.*

"single-mindedly working": Gomory and Sylla, "The American Corporation."

the 1970s to the 1980s: Reich, *Saving Capitalism.*

to decrease labor costs: Robert B. Reich, "When Bosses Shared Their Profits," *New York Times,* June 26, 2020, https://www.nytimes.com/2020/06/25/opinion/sunday/corporate-profit-sharing-inequality.html.

95 *"central importance to":* Reich, *Saving Capitalism.*

corporate bottom lines: Dean Baker, "NAFTA Lowered Wages, As It Was Supposed to Do," *New York Times,* December 5, 2013.

U.S. trade negotiations: Lawrence O. Gostin, Neil R. Sircar, and Eric A. Friedman, "How the US Elevates Corporate Interests over Global Public Health. And How the World Can Respond," *Health Affairs Blog,* September 5, 2018, https://www.healthaffairs.org/do/10.1377/hblog20180830.186562/full/.

in the mid-1950s: Steven Greenhouse, "Union Membership in U.S. Fell to a 70-Year Low Last Year," *New York Times,* January 21, 2011; U.S. Bureau of Labor Statistics, "Union Members Summary, Economic News Release," January 22, 2020, https://www.bls.gov/news.release/union2.nr0.htm.

"The principal objective": Gomory and Sylla, "The American Corporation."

96 *CEO compensation:* Peter Eavis, "Executive Pay: Invasion of the Supersalaries," *New York Times,* April 12, 2014, http://www.nytimes.com/2014/04/13/business/executive-pay-invasion-of-the-supersalaries.html.

276 times greater: Roger Lowenstein, "CEO Pay Is Out of Control. Here's How to Rein It In," *Fortune,* April 19, 2017, http://fortune.com/2017/04/19/executive-compensation-ceo-pay/.

thirteen other wealthy countries: David Leonhardt and Yaryna Serkez, "The U.S. Is Lagging Behind Many Rich Countries. These Charts Show Why," *New York Times,* July 2, 2020, https://www.nytimes.com/interactive/2020/07/02/opinion/politics/us-economic-social-inequality.html.

had grown to 320: David Gelles, "C.E.O. Pay Remains Stratospheric, Even at Companies Battered by Pandemic," *New York Times,* April 24, 2021, https://www.nytimes.com/2021/04/24/business/ceos-pandemic-compensation.html.

astounding 55 percent: Helaine Olen, "As Wealthy CEOs Rake in Money, an Ugly Trope About Americans Needing Help Reemerges," *Washington Post,* May 19, 2021.

than any other industry: See https://www.opensecrets.org/lobby/top. php.

535 members of Congress: Stuart Silverstein, "This Is Why Your Drug Prescriptions Cost So Damn Much," *Mother Jones,* October 21, 2016, https://www.motherjones.com/politics/2016/10/drug-industry-phar maceutical-lobbyists-medicare-part-d-prices/.

each member of Congress: Center for Responsive Politics, "Industry Profile: Pharmaceuticals/Health Products," OpenSecrets.org, https:// www.opensecrets.org/federal-lobbying/industries/summary?cycle =2019&id=h04.

97   *risk associated with Vioxx:* Richard Horton, "Vioxx, the Implosion of Merck, and Aftershocks at the FDA," *Lancet* 364 (2004): 1995–96.

estimated tens of thousands: Graham et al., "Risk of Acute Myocardial Infarction."

opted for early retirement: Raymond Gilmartin, "Merck CEO Gilmartin Steps Down," interview by Joanne Silberner, *Morning Edition,* NPR, May 6, 2005, https://www.npr.org/templates/story/story.php?storyId= 4632943.

"descent into a morass": Christopher Bowe, "Merck's Fall from Grace," *Financial Times,* November 17, 2004, https://www.ft.com/content/581c dc6e-38d2-11d9-bc76-00000e2511c8.

rarely jailed: Sammy Almashat et al., "Twenty-Seven Years of Pharmaceutical Industry Criminal and Civil Penalties: 1991 Through 2017," *Public Citizen,* www.citizen.org, March 14, 2018, https://www.citizen.org/ wp-content/uploads/2408.pdf.

thirty-five years after: "Corporate Profits Versus Labor Income: Risk and Reward Versus Slow and Steady," *FRED Blog,* Federal Reserve Bank of St. Louis, August 9, 2018, https://fredblog.stlouisfed.org/2018/08/corpo rate-profits-versus-labor-income/.

greatest income growth: David Leonhardt, "Our Broken Economy, in One Simple Chart," *New York Times,* August 7, 2017, https://www.nytimes. com/interactive/2017/08/07/opinion/leonhardt-income-inequality .html.

98   *the top 1 percent:* See https://www.census.gov/data/tables/time-series/ demo/income-poverty/historical-income-households.html.

versus 242 percent: Chad Stone et al., "A Guide to Statistics on Historical Trends in Income Inequality," Center on Budget and Policy Prior-

ities, January 13, 2020, https://www.cbpp.org/sites/default/files/atoms/
files/11-28-11pov_0.pdf.

the wealth gap: Thomas Shapiro, Tatjana Meschede, and Sam Osoro,
"The Roots of the Widening Racial Wealth Gap: Explaining the Black-
White Economic Divide," Institute on Assets and Social Policy, https://
heller.brandeis.edu/iasp/pdfs/racial-wealth-equity/racial-wealth-gap/
roots-widening-racial-wealth-gap.pdf.

without *college degrees:* Christian Weller, "African-Americans' Wealth a
Fraction of That of Whites Due to Systematic Inequality," *Forbes,* February 14, 2019, https://www.forbes.com/sites/christianweller/2019/02/14/
african-americans-wealth-a-fraction-that-of-whites-due-to-systematic-
inequality/#1a8414084554.

least equal distributions: "Income Inequality," OECD, oecd-ilibrary.org,
doi: 10.1787/459aa7f1-en.

highest poverty rates: "Poverty Rate," OECD, data.oecd.org, https://data.
oecd.org/ inequality/poverty-rate.htm.

fifty years earlier: Thomas Piketty, *Capital and Ideology,* trans. Arthur
Goldhammer (Cambridge, MA: Harvard University Press, 2019), 530.

ten other wealthy countries: Robin Osborn et al., "In New Survey of 11
Countries, U.S. Adults Still Struggle with Access to and Affordability of Health Care," Commonwealth Fund, November 16, 2016, https://
www.commonwealthfund.org/publications/journal-article/2016/nov/
new-survey-11-countries-us-adults-still-struggle-access-and?redirect_
source=/publications/in-the-literature/2016/nov/2016-international-
health-policy-survey-of-adults.

less than $55,775: Kirby C. Posey, "Household Income, 2015: Ameri-
can Community Survey Briefs," U.S. Census Bureau, September 2016,
https://census.gov/content/dam/Census/library/publications/2016/acs/
acsbr15-02.pdf.

to the 1 percent: Piketty, *Capital and Ideology,* 524.

$480,000 per year: Joseph Mysak Jr., "To Get into the 1%, You Need Ad-
justed Gross Income of $480,930," Bloomberg.com, February 22, 2018,
https://www.bloomberg.com/news/articles/2018-02-22/to-get-into-
the-1-you-need-adjusted-gross-income-of-480-930.

99   *"What primarily characterizes":* Thomas Piketty, *Capital in the Twen-
ty-First Century,* trans. Arthur Goldhammer (Cambridge, MA: Harvard
University Press, 2017), 232.

maintains the inequality: Piketty, *Capital and Ideology,* 2.

in every state in the country: Sara R. Collins, David C. Radley, and
Jesse C. Baumgartner, "Trends in Employer Health Care Coverage,

2008–2018: Higher Costs for Workers and Their Families," Commonwealth Fund, November 21, 2019, https://www.commonwealthfund.
org/publications/2019/nov/trends-employer-health-care-coverage-2008
-2018.

(down by 43 percent): Don Berwick, "Hit the Brakes on the Partners
Deal," *Boston Globe,* October 18, 2014, https://www.bostonglobe.com/
opinion/editorials/2014/10/18/hit-brakes-partners-deal/F6KHBP
3wiIXJZHUPabd8jL/story.html.

101   *than for white Americans:* National Center for Health Statistics, Centers
for Disease Control and Prevention, July 26, 2018, https://www.cdc.gov/
nchs/data/nvsr/nvsr67/nvsr67_05.pdf.

who graduated from college: Anne Case and Angus Deaton, *Deaths of Despair and the Future of Capitalism* (Princeton, NJ: Princeton University
Press), 57.

between 1998 and 2015: Case and Deaton, "Mortality and Morbidity,"
416.

especially cardiovascular disease: Author's personal communication
with Angus Deaton, November 30, 2020.

102   *"the failure of life":* Case and Deaton, "Mortality and Morbidity," 434.

"Warren Buffett likened": Case and Deaton, *Deaths of Despair,* 187.

"It is questionable": Victor R. Fuchs, "Is Single Payer the Answer for the
US Health Care System?," *Journal of the American Medical Association*
319, no. 1 (2018): 15–16, doi:10.1001/jama.2017.18739.

6. HOW DOCTORS KNOW

104   *"The social acceptance":* Steve Fuller, *Social Epistemology* (Bloomington:
Indiana University Press, 1988), 10.

spending on prescription drugs: Ezekiel J. Emanuel, "The Real Cost of the
US Health Care System," *Journal of the American Medical Association* 319,
no. 10 (2018): 983–85.

other wealthy nations: Dana O. Sarnak, David Squires, and Shawn
Bishop, "Paying for Prescription Drugs Around the World: Why Is the
U.S. an Outlier?," Commonwealth Fund, October 5, 2017, https://www.
commonwealthfund.org/publications/issue-briefs/2017/oct/paying-pre
scription-drugs-around-world-why-us-outlier.

brand-name drugs: Aaron S. Kesselheim, Jerry Avorn, and Ameet Sar-
patwari, "The High Cost of Prescription Drugs in the United States: Origins and Prospects for Reform," *Journal of the American Medical Association* 316, no. 8 (2016): 858–71.

105 *justifies their higher cost:* Sarnak, Squires, and Bishop, "Paying for Prescription Drugs."

set their own prices: David Blumenthal, "It's the Monopolies, Stupid!," *Commonwealth Fund,* May 24, 2018, https://www.commonwealthfund.org/blog/2018/its-monopolies-stupid.

the other OECD countries: Andrew W. Mulcahy, Christopher M. Whaley, Mahlet Gizaw et al., "International Prescription Drug Price Comparisons: Current Empirical Estimates and Comparisons with Previous Studies," *Rand Corporation,* 2021.

"It's the prices, stupid": Blumenthal, "It's the Monopolies."

ten times as much: Joel Lexchin, "Affordable Biologics for All," *JAMA Network Open* 3, no. 4 (2020): e204753.

106 *unavailable to peer reviewers:* Roni Caryn Rabin, "The Pandemic Claims New Victims: Prestigious Medical Journals," *New York Times,* June 14, 2020, https://www.nytimes.com/2020/06/14/health/virus-journals.html.

108 *these external factors:* Robert M. Kaplan and Arnold Milstein, "Contributions of Health Care to Longevity: A Review of 4 Estimation Methods," *Annals of Family Medicine* 17, no. 3 (2019): 267–72, https://doi.org/10.1370/afm.2362.

109 *fallen to about one-third:* Office of Extramural Research, "Actual of NIH R01 Equivalent Success Rate: FY 1970–2008 and Estimates of NIH R01," report 1649, National Institutes of Health, 2008.

drug industry and medical leaders: Dominique Tobbell, *Pills, Power, and Policy: The Struggle for Drug Reform in Cold War America and Its Consequences* (Berkeley: University of California Press, 2012), 182–83.

110 *especially the pharmaceutical industry:* B. J. Culliton, "The Academic-Industrial Complex," *Science* 216 (1982): 960–62.

had supported the legislation: Daniel M. Fox, "Policy Commercializing Nonprofits in Health: The History of a Paradox from the 19th Century to the ACA," *Milbank Quarterly* 93, no. 1 (2015): 179–210.

"an uneasy sense that": Culliton, "The Academic-Industrial Complex."

"concern about whether": Culliton, "The Academic-Industrial Complex."

commercial research studies: "Careers in Clinical Research: Obstacles and Opportunities (1994)," *Consensus Study Report* (Washington, DC: National Academies Press, 1994), 206.

receiving industry money: David Blumenthal et al., "Participation of Life-Science Faculty in Research Relationships with Industry," *New England Journal of Medicine* 335 (1996): 1734–39.

therapy for their patients: Susan F. Wood et al., "Influence of Pharmaceutical Marketing on Medicare Prescriptions in the District of Columbia," *PLoS ONE* 12, no. 10 (October 25, 2017): e0186060, https://doi.org/10.1371/journal.pone.0186060.

111 *drug company logos:* Thanks to Isabel Goldman for this wonderful image.

maximize the financial return: Thomas Bodenheimer, "Uneasy Alliance — Clinical Investigators and the Pharmaceutical Industry," *New England Journal of Medicine* 342 (2000): 1539–44.

"a race to the ethical bottom": Drummond Rennie, "Thyroid Storm," *Journal of the American Medical Association* 277 (1997): 1238–43.

112 *"Their intention is to":* Melody Petersen, "Madison Ave. Has Growing Role in the Business of Drug Research," *New York Times,* November 22, 2002.

"a mockery of clinical": Frank Davidoff et al., "Sponsorship, Authorship, and Accountability," *New England Journal of Medicine* 345 (2001): 825–27.

80 to 26 percent: Robert Steinbrook, "Gag Clauses in Clinical-Trial Agreements," *New England Journal of Medicine* 352, no. 21 (2005): 2160–62.

113 *"own the data produced":* Michelle M. Mello, Brian R. Clarridge, and David M. Studdert, "Academic Medical Centers' Standards for Clinical-Trial Agreements with Industry," *New England Journal of Medicine* 352 (2005): 2202–10.

from 23 to 14 percent: Kathryn Doyle, "NIH-Funded Trials Dip, Industry Trials on the Rise," *Scientific American,* December 16, 2015, https://www.scientificamerican.com/article/nih-funded-trials-dip-industry-trials-on-the-rise/.

114 *absolutely no reduction:* D. Y. Graham, N. P. Jewell, and F.K.L. Chan, "Rofecoxib and Clinically Significant Upper and Lower GI Events Revisited Based on Documents from Recent Litigation," *American Journal of the Medical Sciences* 342, no. 5 (November 2011): 356–64, doi:10.1097/MAJ.0b013e3182113658.

excluded from participation: Heart Protection Study Collaborative Group, "MRC/BHF Heart Protection Study of Cholesterol Lowering with Simvastatin in 20,536 High-Risk Individuals: A Randomised Placebo-Controlled Trial," *Lancet* 360 (2002): 7–22.

115 *conventional anti-inflammatory drugs:* Two studies, one published in 1996 in the *New England Journal of Medicine* and the other in 1999 in *The Lancet,* showed that triple therapy was highly effective when one or two of the three drugs taken individually had failed. See James R. O'Dell et al., "Treatment of Rheumatoid Arthritis with Methotrexate Alone, Sul-

fasalazine and Hydroxychloroquine, or a Combination of All Three Medications," *New England Journal of Medicine* 334 (1996): 1287–91, and Timo
Möttönen et al., "Comparison of Combination Therapy with Single-Drug
Therapy in Early Rheumatoid Arthritis: A Randomised Trial," *Lancet* 353
(1999): 1568–73.

for the treatment of RA: Glen S. Hazlewood et al., "Methotrexate Mono-
therapy and Methotrexate Combination Therapy with Traditional and
Biologic Disease Modifying Antirheumatic Drugs for Rheumatoid Arthritis: Abridged Cochrane Systematic Review and Network Meta-Analysis," *British Medical Journal* 353 (2016): i1777.

provide adequate relief: Nick Bansback et al., "Triple Therapy Versus Biologic Therapy for Active Rheumatoid Arthritis: A Cost-Effectiveness
Analysis," *Annals of Internal Medicine* 167 (2017): 8–16, doi:10.7326/M16-
0713.

116   *Pfizer's view of data ownership:* Glen Spielmans and Peter Parry, "From
Evidence-Based Medicine to Marketing-Based Medicine: Evidence from
Internal Industry Documents." Copyright © 2010 by Elsevier Ltd. Re-
printed with permission from Elsevier.

not just boilerplate: Glen I. Spielmans and Peter I. Parry, "From Evidence-Based Medicine to Marketing-Based Medicine: Evidence from Internal Industry Documents," *Journal of Bioethical Inquiry* 7 (2010): 13–29.

117   *risks of the company's drugs:* Mello, Clarridge, and Studdert, "Academic
Medical Centers' Standards."

118   *or used for marketing:* Adapted from Peter Doshi et al., "Restoring Invisible and Abandoned Trials: A Call for People to Publish the Findings,"
British Medical Journal 346 (2013): f2865.

119   *(by academic researchers):* Kristine Rasmussen et al., "Collaboration
Between Academics and Industry in Clinical Trials: Cross Sectional
Study of Publications and Survey of Lead Academic Authors," *British Medical Journal* 363 (2018): k3654, http://dx.doi.org/10.1136/bmj
.k3654.

several thousand pages: Integrated Addendum to ICH E6(R1): "Guideline for Good Clinical Practice E6(R2)," Current Step 4 Version, November 9, 2016, http://www.ich.org/fileadmin/Public_Web_Site/ ICH_Prod
ucts/Guidelines/Efficacy/E6/E6_R2__Step_4_2016_1109.pdf.

120   *access to the underlying data:* Rasmussen et al., "Collaboration."

121   *"a race to the ethical bottom":* Rennie, "Thyroid Storm."

optimal patient care for: Gordon Guyatt et al., "Evidence-Based Medicine: A New Approach to Teaching the Practice of Medicine," *Journal of
the American Medical Association* 268, no. 17 (1992): 2420–25.

122 *"trustworthy clinical practice":* Benjamin Djulbegovic and Gordon H. Guyatt, "Progress in Evidence-Based Medicine: A Quarter Century On," *Lancet* 390 (2017): 415–23.

about optimal patient care: Trisha Greenhalgh, Jeremy Howick, and Neal Maskrey, "Evidence Based Medicine: A Movement in Crisis?," *British Medical Journal* 348 (2014): g3725.

123 *"Evidence without data":* "Medicine's Most Infamous Clinical Trial," Restoring Study 329, https://study329.org/.

124 *the practice of medicine:* Djulbegovic and Guyatt, "Progress in Evidence-Based."

six out of seven: Doyle, "NIH-Funded Trials."

34 percent more frequently: Andreas Lundh et al., "Industry Sponsorship and Research Outcome," *Cochrane Database of Systematic Reviews* 2 (2017): MR000033.

the drug is effective: Rose Ahn et al., "Financial Ties of Principal Investigators and Randomized Controlled Trial Outcomes: Cross Sectional Study," *British Medical Journal* 356 (2017): i6770, http://dx.doi.org/10.1136/bmj.i6770.

funded by industry: Nikolaos A. Patsopoulos, John P. A. Ioannidis, and Apostolos A. Analatos, "Origin and Funding of the Most Frequently Cited Papers in Medicine: Database Analysis," *British Medical Journal* 332 (2006): 1061–64.

below-the-waterline data: Beate Wieseler et al., "Completeness of Reporting of Patient-Relevant Clinical Trial Outcomes: Comparison of Unpublished Clinical Study Reports with Publicly Available Data," *PLoS Medicine* 10 (2013): e1001526.

before the trial began: Christopher W. Jones et al., "Primary Outcome Switching Among Drug Trials With and Without Principal Investigator Financial Ties to Industry: A Cross-Sectional Study," *British Medical Journal Open* 8 (2018): e019831.

125 *only about 50 percent:* Beate Wieseler et al., "Impact of Document Type on Reporting Quality of Clinical Drug Trials: A Comparison of Registry Reports, Clinical Study Reports, and Journal Publications," *British Medical Journal* 344, no. 9 (2011): d8141, https://www.bmj.com/content/bmj/344/bmj.d8141.full.pdf.

126 *increased by 2,500 percent:* John Ioannidis, "We Have an Epidemic of Deeply Flawed Meta-Analyses, Says John Ioannidis," interview with Retraction Watch, September 13, 2016, https://retractionwatch.com/2016/09/13/we-have-an-epidemic-of-deeply-flawed-meta-analyses-says-john-ioannidis/.

most frequently cited type: Djulbegovic and Guyatt, "Progress in Evidence-Based."

the drug being reviewed: Shanil Ebrahim et al., "Meta-Analyses with Industry Involvement Are Massively Published and Report No Caveats for Antidepressants," *Journal of Clinical Epidemiology* 70 (February 2016): 155–63.

40 percent and 22 percent: Anders W. Jørgensen et al., "Industry-Supported Meta-Analyses Compared with Meta-Analyses with Non-Profit or No Support: Differences in Methodological Quality and Conclusions," *BMC Medical Research Methodology* 8 (2008): 60, https://doi.org/10.1186/1471-2288-8-60.

"decent and clinically useful": John P. A. Ioannidis, "The Mass Production of Redundant, Misleading, and Conflicted Systematic Reviews and Meta-Analyses," *Milbank Quarterly* 94 (2016): 485–514.

127   *the Cochrane Collaboration:* Djulbegovic and Guyatt, "Progress in Evidence-Based."

"to produce high-quality": Cochrane, https://www.cochrane.org/about-us/strategy-to-2020.

acute flu-like symptoms: Peter Doshi, "Neuraminidase Inhibitors — the Story Behind the Cochrane Review," *British Medical Journal* 339 (2009): 1348–51; Laurent Kaiser et al., "Impact of Oseltamivir Treatment on Influenza-Related Lower Respiratory Tract Complications and Hospitalizations," *Archives of Internal Medicine* 163 (2003): 1667–72.

"in a serious epidemic": T. O. Jefferson et al., "Neuraminidase Inhibitors for Preventing and Treating Influenza in Healthy Adults," *Cochrane Database of Systematic Reviews* 3 (2006): CD001265.

forty million doses: Associated Press, "Drug Manufacturer Says U.S. Wants to Increase Tamiflu Stockpile," *Fox News,* January 31, 2006, updated January 13, 2015, www.foxnews.com/story/2006/01/31/drug-manufacturer-says-us-wants-to-increase-tamiflu-stockpile.html.

essential or "core": Michael McCarthy, "What Makes an Essential Medicine? WHO's New List Focuses on Antibiotic Resistance, Adds Expensive Drugs, and Downgrades Tamiflu," *British Medical Journal* 358 (2017): j3044.

could not be verified: Mark A. Jones et al., "Commentary on Cochrane Review of Neuraminidase Inhibitors for Preventing and Treating Influenza in Healthy Adults and Children," *Clinical Microbiology and Infection* 21 (2015): 217–21.

128   *grossly inadequate and evasive:* Doshi, "Neuraminidase Inhibitors."

seventy-seven trials of Tamiflu: Tom Jefferson and Peter Doshi, "Multisystem Failure: The Story of Anti-Influenza Drugs," *British Medical Journal* 348 (2014): g2263.

Roche's bottom line: Andrew Jack, "Tamiflu: 'A Nice Little Earner,'" *British Medical Journal* 348 (2014): g2524.

less cost effective: Zosia Kmietowicz, "WHO Downgrades Oseltamivir on Drugs List After Reviewing Evidence," *British Medical Journal* 357 (2017): j2841.

by governments for stockpiling: Mark H. Ebell, "WHO Downgrades Status of Oseltamivir," *British Medical Journal* 358 (2017): j3266.

129 *limited time and resources:* Alex Hodkinson et al., "The Use of Clinical Study Reports to Enhance the Quality of Systematic Reviews: A Survey of Systematic Review Authors," *Systematic Reviews* 7 (2018), https://doi.org/10.1186/s13643-018-0766-x.

evidence-based medicine: Djulbegovic and Guyatt, "Progress in Evidence-Based."

"ceaselessly escalating": Institute of Medicine, *Clinical Practice Guidelines We Can Trust* (Washington, DC: National Academies Press, 2011), 16.

IOM issued a report: Institute of Medicine, *Clinical Practice,* 17.

met less than half: Terrence M. Shaneyfelt, Michael F. Mayo-Smith, and Johann Rothwangl, "Are Guidelines Following Guidelines? The Methodological Quality of Clinical Practice Guidelines in the Peer-Reviewed Medical Literature," *Journal of the American Medical Association* 281 (1999): 1900–1905.

its standards for guidelines: Institute of Medicine, *Clinical Practice,* 20.

130 *"demonstrated poor compliance":* Justin Kung, Ram R. Miller, and Phillip A. Mackowiak, "Failure of Clinical Practice Guidelines to Meet Institute of Medicine Standards: Two More Decades of Little, If Any, Progress," *Archives of Internal Medicine* 172, no. 21 (2012): 1628–33.

seven times more likely: Ariadna Tibau et al., "Author Financial Conflicts of Interest, Industry Funding, and Clinical Practice Guidelines for Anticancer Drugs," *Journal of Clinical Oncology* 33 (2014): 100–106.

"the extent to which clinical": "Guidelines and Measures Summaries," AHRQ, https://www.guideline.gov/help-and-about/summaries/determining-extent-adherence-to-the-iom-standards-in-ngc.

AHRQ defunded it: Ivan Oransky and Adam Marcus, "Trump Administration Is Shutting Down Practice-Guidelines Clearinghouse for Doctors," *STAT News,* June 13, 2018, https://www.statnews.com/2018/06/13/ahrq-practice-guidelines-clearinghouse-shutting-down/.

the Wayback Machine: National Guideline Clearinghouse, https://web.

archive.org/web/20180609103505/https://www.guideline.gov/help-and-about/summaries/determining-extent-adherence-to-the-iom-standards-in-ngc.

131   *ten "high-revenue" drugs:* Rishad Khan et al., "Prevalence of Financial Conflicts of Interest Among Authors of Clinical Guidelines Related to High-Revenue Medications," Research Letter, *JAMA Internal Medicine* 178, no. 12 (2018): 1712–15.

only 34 percent had declared: Tyler R. Combs et al., "Evaluation of Industry Relationships Among Authors of Clinical Practice Guidelines in Gastroenterology," Research Letter, *JAMA Internal Medicine* 178, no. 12 (2018): 1711–12.

"If the specialty societies": Colette DeJong and Robert Steinbrook, "Continuing Problems with Financial Conflicts of Interest and Clinical Practice Guidelines," *JAMA Internal Medicine* 178, no. 12 (2018): 1715.

"Pharmaceutical and biotechnology": "American College of Rheumatology Principles Regarding External Entity Support for Rheumatology Fellowship Training," ACR Committee on Rheumatology Training and Workforce Issues, January 2015, 1–5, https://www.rheumatology.org/Portals/0/Files/Principles-Governing-Industry-Support-for-Rheumatology-Fellowship-Training.pdf.

"It is difficult to get": Upton Sinclair, *I, Candidate for Governor: And How I Got Licked* (Berkeley: University of California Press, 1934), 109.

132   *"reproducibility, rigor":* Marcia McNutt, "Journals Unite for Reproducibility," *Science* 346, no. 6 (2014): 679.

133   *"facts determined by experiment":* "History of the Royal Society," royalsociety.org, https://royalsociety.org/about-us/history/.

rather than a public good: Mark A. Rodwin and John D. Abramson, "Clinical Trial Data as a Public Good," *Journal of the American Medical Association* 308, no. 9 (2012): 871–72.

7. MANUFACTURING BELIEF

134   *"Modern economics inherently":* George A. Akerlof and Robert J. Shiller, *Phishing for Phools: The Economics of Manipulation and Deception* (Princeton, NJ: Princeton University Press, 2015), 164.

135   *"substantially equivalent":* Brent M. Ardaugh, Stephen E. Graves, and Rita F. Redberg, "The 510(k) Ancestry of a Metal-on-Metal Hip Implant," *New England Journal of Medicine* 368 (2013): 97–100.

136   *"overweight or very active":* "In Re: DePuy Orthopedics, Inc. Pinnacle Hip Implant Products Liability Litigation," vol. 31, transcript of trial be-

fore the Honorable Ed Kinkeade, United States district judge, and a jury, November 30, 2016, 213.

trial went forward: Reuters, "Johnson & Johnson Just Got Hit with a $1B Verdict over Faulty Hip Implants," *Fortune,* December 2, 2016, https://fortune.com/2016/12/02/johnson-johnson-just-got-hit-with-a-1b-verdict-over-faulty-hip-implants/.

137 *Pinnacle hip implants:* Joan E. Steffen et al., "Grave Fraudulence in Medical Device Research: A Narrative Review of the PIN Seeding Study for the Pinnacle Hip System," *Accountability in Research* 25 (2018): 37–66.

138 *"fundamental selling point":* Steffen et al., "Grave Fraudulence."
French regulatory authority: Steffen et al., "Grave Fraudulence."
metal-on-polyethylene implants: Steffen et al., "Grave Fraudulence."
"prejudicial effect": "In Re: DePuy Orthopedics, Inc.," vol. 4, October 5, 2016, 188–94.

140 *$1.04* billion *award:* "Punitive Damages," Legal Information Institute, https://www.law.cornell.edu/wex/punitive_damages.
(details of the final settlement): "DePuy ASR & Pinnacle Hip Replacement Lawsuits," Drugwatch, https://www.drugwatch.com/hip-replacement/depuy/lawsuits/.
legal and moral obligation: Cara D. Edwards and Brooke Killian Kim, "The Learned Intermediary Doctrine in the WebMD Era," *DLA Piper,* August 1, 2019, https://www.dlapiper.com/en/us/insights/ publications/2019/06/the-learned-intermediary-doctrine-in-the-webmd -era/.

141 *safety and efficacy:* Ardaugh et al., "The 510(k) Ancestry of a Metal-on-Metal Hip Implant."
eleven times more likely: Charles S. Day et al., "Analysis of FDA-Approved Orthopaedic Devices and Their Recalls," *Bone and Joint Surgery* 98, no. 6 (2016): 517–24, http://dx.doi.org/10.2106/JBJS.15.00286.
"novel drugs — innovative": "Drug Industry: Profits, Research and Development Spending, and Merger and Acquisition Deals," Report to Congressional Requesters, U.S. Government Accountability Office, November 2017, https://www.gao.gov/assets/690/688472.pdf.

142 *withholding or misrepresenting:* Aidan R. Vining and David L. Weimer, "Information Asymmetry Favoring Sellers: A Policy Framework," *Policy Sciences* 21, no. 4 (1988): 281–303. What I am calling a "credence good," Vining and Weimer refer to as a "post-experience good."
in effect, mandates: Stephen Bainbridge, "A Duty to Shareholder Value," *New York Times,* April 16, 2015.

143   *accepted as knowledge:* Susan Haack, *Evidence and Inquiry: Towards Reconstruction in Epistemology* (Oxford: Wiley-Blackwell, 1995), 206.
Their book debunks: Akerlof and Shiller, *Phishing,* Kindle location 429.

144   *aim about two-thirds:* Lisa M. Schwartz and Steven Woloshin, "Medical Marketing in the United States, 1997–2016," *Journal of the American Medical Association* 321, no. 1 (2019): 80–96.
"an extremely good": Author's personal communication with George A. Akerlof, November 30, 2013.
"gaming the doctors": Akerlof and Shiller, *Phishing,* 92–93.
These behaviors: Akerlof and Shiller, *Phishing,* 7.

145   *accepted by 70 percent:* Robert Steinbrook, "Physicians, Industry Payments for Food and Beverages, and Drug Prescribing," *Journal of the American Medical Association* 317, no. 17 (2017): 1753–54.
increases doctors' prescribing: Colette DeJong et al., "Pharmaceutical Industry–Sponsored Meals and Physician Prescribing Patterns for Medicare Beneficiaries," *JAMA Internal Medicine* 176, no. 8 (2016): 1114–22.
effect on physicians' prescribing: Aaron P. Mitchell et al., "Are Financial Payments from the Pharmaceutical Industry Associated with Physician Prescribing? A Systematic Review," *Annals of Internal Medicine* (November 2020), https://www.acpjournals.org/doi/10.7326/M20-5665.
Drug reps are trained: Shahram Ahari, "I Was a Drug Rep. I Know How Pharma Companies Pushed Opioids," *Washington Post,* November 26, 2019, https://www.washingtonpost.com/outlook/i-was-a-drug-rep-i-know-how-pharma-companies-pushed-opioids/2019/11/25/82b1da88-beb9-11e9-9b73-fd3c65ef8f9c_story.html.

146   *fulfillment of marketers' goals:* Akerlof and Shiller, *Phishing,* 52.
"act on information that": Akerlof and Shiller, *Phishing,* xi.

147   *effective phishing of doctors:* Abramson, "Expert Report."
evidence delivered to physicians: Kaiser v. Pfizer, September 2002, trial exhibit 259.

149   *benefits of their products:* Merve Memişoğlu, "Branding of Prescription and Non-Prescription Drugs," *Acta Pharmaceutica Sciencia* 56 (2018): 21–36.
$6 billion in annual sales: James Gerard Conley, Robert C. Wolcott, and Eric Wong, "AstraZeneca, Prilosec, and Nexium: Marketing Challenges in the Launch of a Second-Generation Drug," *Kellogg Case Publishing,* April 1, 2005, http://web.archive.org/web/20180301012750/https://www.kellogg.northwestern.edu/kellogg-case-publishing/case-search/case-detail.aspx?caseid=%7B34C00D56-6CC6-4598-B321-8D3F61E0AAF7%7D.

best-selling drug in the world: "Best-Selling Prescription Drug Aims to Go over the Counter," WebMD, October 16, 2000, https://www.webmd.com/drug-medication/news/20001016/best-selling-prescription-drug-aims-to-go-over-counter#1.

replacement for Prilosec: Gardner Harris, "Fast Relief: As a Patent Expires, Drug Firm Lines Up Pricey Alternative," *Wall Street Journal,* June 6, 2002.

"among the poorest": Harris, "Fast Relief."

"a significant clinical advance": "21154 Nexium Medical Review Part 5," New Drug Application 21-153, submitted to the Federal Drug Administration February 28, 2000, 171, https://www.accessdata.fda.gov/drug satfda_docs/nda/2001/21154_Nexium_medr_P5.pdf.

"You should be embarrassed": Gardner Harris, "2 New Fronts in Heartburn Market Battle," *New York Times,* August 20, 2003, https://www.nytimes.com/2003/08/20/business/2-new-fronts-in-heartburn-market-battle.html.

150 *most commercially successful:* Carolyn Fitzsimons, http://astrazeneca.com/sites/7/archive/Investors/Presentations/2004/astrazeneca-2004-abr-carolyn-fitzsimons-nexium.pdf.

$72 billion in U.S. sales: Angus Liu, "From Old Behemoth Lipitor to New King Humira: Best-Selling U.S. Drugs over 25 Years," *FiercePharma,* May 14, 2018, https://www.fiercepharma.com/pharma/from-old-behemoth-lipitor-to-new-king-humira-u-s-best-selling-drugs-over-25-years.

151 *AZ would have had to:* John Abramson, *Overdo$ed America: The Broken Promise of American Medicine* (New York: HarperCollins, 2004).

evidence of the superiority: "21154 Nexium Medical Review Part I," New Drug Application 21-153/21-154, submitted to the Federal Drug Administration February 28, 2000, 5–6, https://www.accessdata.fda.gov/drug satfda_docs/nda/2001/21154_Nexium_medr_P1.pdf.

published in 2000 and 2001: P. J. Kahrilas et al., "Esomeprazole Improves Healing and Symptom Resolution as Compared with Omeprazole in Reflux Oesophagitis Patients: A Randomized Controlled Trial," *Alimentary Pharmacology and Therapeutics* 14, no. 10 (2000): 1249–58; J. E. Richter et al., "Efficacy and Safety of Esomeprazole Compared with Omeprazole in GERD Patients with Erosive Esophagitis: A Randomized Controlled Trial," *American Journal of Gastroenterology* 96, no. 3 (2001): 656–65.

until 2004 and 2006: D. Armstrong et al., "The Role of Acid Suppression in Patients with Endoscopy-Negative Reflux Disease: The Effect of Treatment with Esomeprazole or Omeprazole," *Alimentary Pharmacology and Therapeutics* 20, no. 4 (2004): 413–21; Colleen Schmitt et al., "A

Multicenter, Randomized, Double-Blind, 8-Week Comparative Trial of Standard Doses of Esomeprazole (40 mg) and Omeprazole (20 mg) for the Treatment of Erosive Esophagitis," *Digestive Diseases and Sciences* 51, no. 5 (2006): 844–50; Charles J. Lightdale et al., "A Multicenter, Randomized, Double-Blind, 8-Week Comparative Trial of Low-Dose Esomeprazole (20 mg) and Standard-Dose Omeprazole (20 mg) in Patients with Erosive Esophagitis," *Digestive Disease and Sciences* 51, no. 5 (2006): 852–57.

152   *"The proof's in the healing":* Harris, "Fast Relief."

    a record $1.08 billion: Matthew Arnold, "FDA Eyes Nexium, Prilosec for Heart Risks with Early Communication," *Medical Marketing and Media,* August 9, 2007, http://www.mmm-online.com/FDA-eyes-Nexium-Prilosec-for-heart-risks-with-ampquotearly-communicationquot/article/24926/.

    impression of the superiority: "Nexium Purple Pill 2002 TV Ad Commercial," YouTube, March 9, 2020, https://www.bing.com/videos/search?q=tv+advertisement+for+nexium&docid=608034315567762514&mid=86CC6B3BFADCD319748286CC6B3BFADCD3197482&view=detail&FORM=VIRE.

153   *developed acid-rebound symptoms:* Christina Reimer et al., "Proton-Pump Inhibitor Therapy Induces Acid-Related Symptoms in Healthy Volunteers After Withdrawal of Therapy," *Gastroenterology* 137, no. 1 (2009): 80–87.

    "chemicals plus information": David Healy, "Trust Me I'm Not a Doctor," *Dr. David Healy* (blog), September 16, 2020, https://davidhealy.org/trust-me-im-not-a-doctor/.

    "fake information": David Healy, "Making Medicine Safer for All of Us," *Dr. David Healy* (blog), December 30, 2018, https://davidhealy.org/making-medicines-safer-for-all-of-us/.

154   *less likely to be abused:* See https://media.defense.gov/2007/May/10/2001711223/-1/-1/1/purdue_frederick_1.pdf.

    primary revenue generator: Harriet Ryan, Lisa Girion, and Scott Glover, "'You Want a Description of Hell?' OxyContin's 12-Hour Problem," *Los Angeles Times,* May 5, 2016, https://www.latimes.com/projects/ oxycontin-part1/.

    met the challenge: David Armstrong, "Purdue's Sackler Embraced Plan to Conceal OxyContin's Strength from Doctors, Sealed Deposition Shows," *STAT News,* February 21, 2019, https://www.statnews.com/2019/02/21/purdue-pharma-richard-sackler-oxycontin-sealed-deposition/.

    Germany back in 1916: Katherin Eban, "OxyContin: Purdue Phar-

ma's Painful Medicine," *Fortune,* November 9, 2011, http://fortune.
com/2011/11/09/oxycontin-purdue-pharmas-painful-medicine/.

less *powerful narcotic:* Armstrong, "Purdue's Sackler."

slowed systemic absorption: Casey Ross, "Purdue's Richard Sackler Proposed Plan to Play Down OxyContin Risks, and Wanted Drug Maker Feared 'Like a Tiger,' Files Show," *STAT News,* December 2, 2019, https://
www.statnews.com/2019/12/02/purdue-richard-sackler-proposed-
plan-play-down-oxycontin-risks/.

given just twice daily: Ryan, Girion, and Glover, "'Description of Hell.'"

155   *unusually aggressive marketing:* Art Van Zee, "The Promotion and Marketing of OxyContin: Commercial Triumph, Public Health Tragedy," *American Journal of Public Health* 99, no. 2 (2009): 221–27. https://www.
ncbi.nlm.nih.gov/pmc/articles/PMC2622774/.

other prescription opioids: Barry Meier, "Origins of an Epidemic: Purdue Pharma Knew Its Opioids Were Widely Abused," *New York Times,*
May 29, 2018, https://www.nytimes.com/2018/05/29/health/purdue-opi
oids-oxycontin.html.

funded over twenty thousand: GAO, "Prescription Drugs: OxyContin Abuse and Diversion and Efforts to Address the Problem," Report to Congressional Requesters, U.S. Government Accountability Office, December 2003, https://www.govinfo.gov/content/pkg/GAORE
PORTS-GAO-04-110/pdf/GAOREPORTS-GAO-04-110.pdf.

American Academy of Pain Management: Celine Gounder, "Who Is Responsible for the Pain Pill Epidemic?," *New Yorker,* November 8, 2013,
https://www.newyorker.com/business/currency/who-is-responsible-
for-the-pain-pill-epidemic.

American Geriatric Society: Verna L. Rose, "Guidelines from the American Geriatric Society Target Management of Chronic Pain in Older Persons," *American Family Physician* 58, no. 5 (1998): 1213–15, https://www.
aafp.org/afp/1998/1001/p1213.html.

vital sign in relevant patients: American Pain Society Quality of Care Committee, "Quality Improvement Guidelines for the Treatment of Acute Pain and Cancer Pain," *Journal of the American Medical Association* 274, no. 23 (1995): 1874–80.

only drug company permitted: GAO, "Prescription Drugs," 8.

"*Granted, a few of us*": Abigail Zuger, "A Doctor's Guide to What to Read on the Opioid Crisis," *New York Times Book Review,* December 17, 2018.

156   *Purdue led the list:* HSGAC staff, "Fueling an Epidemic," U.S. Senate Homeland Security and Governmental Affairs Committee, https://www.
hsgac.senate.gov/imo/media/doc/REPORT-Fueling%20an%20Epidem

ic-Exposing%20the%20Financial%20Ties%20Between%20Opioid%20
Manufacturers%20and%20Third%20Party%20Advocacy%20Groups.
pdf.

sales reaching $1 billion: HSGAC staff, "Fueling an Epidemic."

actually twice as strong: David Armstrong, "Sackler Embraced Plan
to Conceal OxyContin's Strength from Doctors, Sealed Testimony
Shows," ProPublica, February 21, 2019, https://www.propublica.org/ar
ticle/richard-sackler-oxycontin-oxycodone-strength-conceal-from-doc
tors-sealed-testimony.

less likely than other narcotics: "Purdue Statement in Response to Leak
of 2015 Deposition of Dr. Richard Sackler," https://www.document
cloud.org/documents/5745864-Purdue-Statement-in-Response-to-
Leak-of-2015.html, 2.

"Delayed absorption, as provided": Marcia Angell, "Opioid Nation," *New
York Review of Books,* December 6, 2018.

similar to that of heroin: GAO, "Prescription Drugs," 2.

for intravenous injection: John Brownlee, "Statement of United States
Attorney John Brownlee on the Guilty Plea of the Purdue Frederick
Company and Its Executives for Illegally Misbranding OxyContin," U.S.
Department of Justice, May 10, 2007, 6, https://www.documentcloud.
org/documents/279028-purdue-guilty-plea.

"that 100% of the patients": Ryan, Girion, and Glover, "'Description of
Hell.'"

needed additional medication: Ryan, Girion, and Glover, "'Description of
Hell.'"

157 *reasons to be untruthful:* Ryan, Girion, and Glover, "'Description of
Hell.'"

"perfect recipe for addiction": Ryan, Girion, and Glover, "'Description of
Hell.'"

"with the intent to defraud": Brownlee, "Statement."

158 *"'blessed' him to do whatever":* Geoff Mulvihill, "OxyContin Maker Purdue Pharma Pleads Guilty in Criminal Case," Associated Press, November 24, 2020.

sales grew to $3 billion: Ryan, Girion, and Glover, "'Description of
Hell.'"

about 14,000 in 2010: "Overdose Death Rates," National Institute on
Drug Abuse, drugabuse.gov, https://www.drugabuse.gov/drug-topics/
trends-statistics/overdose-death-rates.

all the way to 2030: Jessica Hill, "Revamped OxyContin Was Supposed
to Reduce Abuse, but Has It?," *STAT News,* July 22, 2019, https://www.

statnews.com/2019/07/22/revamped-oxycontin-was-supposed-to-reduce-abuse-but-has-it/.

pills sold on the street: George Serletis, "Deadly High-Purity Fentanyl from China Is Entering the U.S. Through E-Commerce Channels," Executive Briefings on Trade, U.S. International Trade Commission, September 2019, https://www.usitc.gov/publications/332/executive_briefings/ebot_george_serletis_fentanyl_from_china_pdf.pdf.

more than 36,000 in 2019: "Opioids: Fentanyl," Centers for Disease Control and Prevention, https://www.cdc.gov/opioids/basics/fentanyl.html.

between 1999 and 2018: "Understanding the Epidemic," CDC, cdc.gov/drugoverdose, https://www.cdc.gov/drugoverdose/epidemic/index.html.

159 *"massive amounts of money":* Chris McGreal, "Big Pharma's Response to the Opioid Epidemic: Pay but Deny," *New York Review,* November 11, 2019, https://www.nybooks.com/daily/2019/11/11/big-pharmas-response-to-the-opioid-epidemic-pay-but-deny/.

tenfold increase in prescriptions: McGreal, "Big Pharma's Response."

"I think that the vast amounts": McGreal, "Big Pharma's Response."

8. MARKET FAILURE IN GENERAL KNOWLEDGE

160 *"Information issues":* John Cassidy, *How Markets Fail: The Logic of Economic Calamities* (New York: Farrar, Straus and Giroux), 163 (Kindle edition). Cassidy is referring to the writing of Nobel laureate Joseph Stiglitz.

"The case against [biomedical] science": Richard Horton, "Offline: What Is Medicine's 5 Sigma?," *Lancet* 385 (2015): 1380.

161 *misleading epidemiological studies:* Francine Grodstein et al., "Postmenopausal Hormone Therapy and Mortality," *New England Journal of Medicine* 336, no. 25 (1997): 1769–76.

increase life expectancy: American College of Physicians, "Guidelines for Counseling Postmenopausal Women About Preventive Hormone Therapy," *Annals of Internal Medicine* 117 (1992): 1038–41.

"Breasts and genital organs": Robert A. Wilson and Thelma A. Wilson, "The Basic Philosophy of Estrogen Maintenance," *Journal of the American Geriatrics Society* 20 (1972): 521–23.

increased *women's risk:* Writing Group for the Women's Health Initiative Investigators, "Risks and Benefits of Estrogen Plus Progestin in Healthy Postmenopausal Women: Principal Results from the Women's Health Initiative Randomized Controlled Trial," *Journal of the American Medical Association* 288, no. 3 (2002): 321–33.

likely to be healthier: Diana Friedman-Koss et al., "The Relationship of Race/Ethnicity and Social Class to Hormone Replacement Therapy: Results from the Third National Health and Nutrition Examination Survey, 1988–1994," *Menopause* 9, no. 4 (2002): 264–72.

162   *scientifically unsubstantiated:* Abramson and Wright, "Are Lipid-Lowering Guidelines."

financial ties to statin makers: Letter from Barbara Alving, acting director of the National Heart, Lung, and Blood Institute, to Merrill Goozner, October 22, 2004, https://web.archive.org/web/20110617175700/http://www.cas.usf.edu/news/response.pdf.

163   *price of drugs is too high:* Rachel Bluth, "Poll: Americans Aghast over Drug Costs but Aren't Holding Their Breath for a Fix," Kaiser Family Foundation, March 23, 2018, https://khn.org/news/poll-americans-aghast-over-drug-costs-but-arent-holding-their-breath-for-a-fix/.

the Consumer Price Index: Aaron S. Kesselheim, Jerry Avorn, and Ameet Sarpatawari, "The High Cost of Prescription Drugs in the United States: Origins and Prospects for Reform," *Journal of the American Medical Association* 316, no. 8 (2016): 858–71.

$1,443 versus $749: Irene Papanicolas, Llana R. Woskie, and Ashish K. Jha, "Health Care Spending in the United States and Other High-Income Countries," *Journal of the American Medical Association* 319, no. 10 (2018): 1024–39.

the only wealthy country: Aaron Kesselheim, personal communication to author, March 5, 2019.

164   *"medication pricing failure":* William H. Shrank, Teresa L. Rogstad, and Natasha Parekh, "Waste in the US Health Care System: Estimated Costs and Potential for Savings," *Journal of the American Medical Association* 322, no. 15 (2019): 1501–9.

cost of drug development: Kesselheim, Avorn, and Sarpatawari, "High Cost."

only a small fraction: Juliette Cubanski et al., "How Does Prescription Drug Spending and Use Compare Across Large Employer Plans, Medicare Part D, and Medicaid?," Medicare and Kaiser Family Foundation, May 20, 2019, https://www.kff.org/medicare/issue-brief/how-does-prescription-drug-spending-and-use-compare-across-large-employer-plans-medicare-part-d-and-medicaid/.

165   *paying 80 percent more:* "The Medicare Drug Price Negotiation Act of 2017: Discussion Draft Summary," oversight.house.gov, https://oversight.house.gov/sites/democrats.oversight.house.gov/files/documents/ Negotiation%20Bill%20Two-Pager%20for%20Release%20-%20Final_0.pdf.

166 *pharmaceutical industry's lobbying:* Paul Blumenthal, "The Legacy of Billy Tauzin: The White House–PhRMA Deal," Sunlight Foundation, February 12, 2010, https://sunlightfoundation.com/ 2010/02/12/the-leg acy-of-billy-tauzin-the-white-house-phrma-deal/.

profit margin on sales: Liyan Chen, "The Most Profitable Industries in 2016," *Forbes,* December 21, 2016, https://www.forbes.com/sites/li yanchen/2015/12/21/the-most-profitable-industries-in-2016/#413b b4ee5716; Bob Herman, "Health Care Profits Explode," Axios, September 19, 2019, https://www.axios.com/health-care-pharma-hospitals-q2-2019-7c20729d-ab9b-460b-9ea8-b08902491eec.html.

as for the S&P 500: Fred D. Ledley et al., "Profitability of Large Pharmaceutical Companies Compared with Other Large Public Companies," *Journal of the American Medical Association* 323, no. 9 (2020): 834–43.

spending on global R&D: Nancy L. Yu, Zachary Helms, and Peter B. Bach, "R&D Costs for Pharmaceutical Companies Do Not Explain Elevated US Drug Prices," *Health Affairs Blog,* March 7, 2017, https://www.healthaffairs.org/do/10.1377/hblog20170307.059036/full/.

167 *extracts a high toll:* Robert Hill and Joseph W. Metro, "OIG Moving Ahead on Changes to Anti-Kickback Safe Harbor Protection for Drug Rebates to Plans, PBMs," Office of Inspector General Regulations, July 20, 2018, https://www.healthindustrywashingtonwatch.com/2018/07/ articles/regulatory-developments/office-of-inspector-general-regula tions/oig-moving-ahead-on-changes-to-anti-kickback-safe-harbor-pro tection-for-drug-rebates-to-plans-pbms/.

168 *living in a different world:* Thomas S. Kuhn, *The Structure of Scientific Revolution* (Chicago: University of Chicago Press, 1962), 192.

169 *"a more systematic approach":* Institute of Medicine, *Knowing What Works in Health Care: A Roadmap for the Nation* (Washington, DC: National Academies Press, 2008), https://www.nap.edu/catalog/12038/ knowing-what-works-in-health-care-a-roadmap-for-the.

170 *"How can our market-driven":* Peter J. Neumann and Milton C. Weinstein, "Legislating Against Use of Cost-Effectiveness Information," *New England Journal of Medicine* 363 (2010): 1495–97.

not to perform *full HTA:* Dana P. Goldman, Samuel Nussbaum, and Mark Linthicum, "Rapid Biomedical Innovation Calls for Similar Innovation in Pricing and Value Measurement," *Health Affairs Blog,* https://www.healthaffairs.org/do/10.1377/hblog20160915.056571 /full/.

171 *excluded from consideration:* "USPSTF and Cost Considerations," U.S. Preventive Services Task Force, February 2017, https://uspreventi

veservicestaskforce.org/uspstf/about-uspstf/task-force-resources/
uspstf-and-cost-considerations.

"I must confess that": Richard Smith, "Medical Journals Are an Exten-
sion of the Marketing Arm of Pharmaceutical Companies," *PLoS Med* 2,
no. 5 (2005): e138, https://doi.org/10.1371/journal.pmed.0020138.

172   The Lancet's *total income*: Andreas Lundh et al., "Conflicts of Interest
at Medical Journals: The Influence of Industry-Supported Randomised
Trials on Journal Impact Factors and Revenue—Cohort Study," *PLoS
Med* (2010): e1000354, doi:10.1371/journal.pmed.1000354.

profits in one year: Richard Smith, "The Hypocrisy of Medical Journals
over Transparency," *BMJ Opinion*, January 24, 2018, https://blogs.bmj.
com/bmj/20a18/01/24/richard-smith-the-hypocrisy-of-medical-jour
nals-over-transparency/.

"likely to be even higher": Smith, "The Hypocrisy."

a whopping 78 percent: Lundh et al., "Conflicts of Interest."

NEJM's *reprint sales*: Smith, "The Hypocrisy."

173   *disclose financial ties*: Rafael Del-Ré et al., "Mandatory Disclosure of Financial Interests of Journals and Editors," *British Medical Journal* 370
(2020): m2872, http://dx.doi.org/10.1136/bmj.m2872.

"checking and rechecking": Horton, "Offline."

redactions would rarely: Barbara Mantel, "Canada Opens the Door to
Public Scrutiny of Clinical Drug Trials," *Undark*, October 9, 2019.

175   *its withdrawal in 2004*: Andrew Pollack, "New Scrutiny of Drugs in Vioxx's Family," *New York Times*, October 4, 2004, https://www.nytimes.
com/2004/10/04/business/new-scrutiny-of-drugs-in-vioxxs-family.
html.

six months after *publication*: Darren B. Taichman et al., "Sharing Clinical Trial Data—a Proposal from the International Committee of Medical
Journal Editors," *New England Journal of Medicine* 374 (2016): 385–86.

"a new class of research": Dan L. Longo and Jeffrey M. Drazen, "Data
Sharing," *New England Journal of Medicine* 374 (2016): 276–77.

176   *"New Science Data-Sharing Rules"*: Adam Marcus and Ivan Oransky,
"New Science Data-Sharing Rules Are Two Scoops of Disappointment,"
STAT News, June 6, 2017, https://www.statnews.com/2017/06/06/da
ta-sharing-rules-disappoint/.

not to share the data: Darren B. Taichman et al., "Data Sharing Statements for Clinical Trials—a Requirement of the International Committee of Medical Journal Editors," *New England Journal of Medicine* 376
(2017): 2277–79, https://doi.org/10.1056/NEJMe1705439.

177   *five-sixths of clinical research*: Stephan Ernhardt, Lawrence J. Appel, and

Curtis L. Meinert, "Trends in National Institutes of Health Funding for Clinical Trials Registered in ClinicalTrials.gov," *Journal of the American Medical Association* 314, no. 23 (2015): 2566–67.

"the health and well-being": Hamilton Moses III et al., "The Anatomy of Medical Research: US and International Comparisons," *Journal of the American Medical Association* 313, no. 2 (2015): 174–89.

Social circumstances and lifestyle: Kaplan and Milstein, "Contributions of Health Care."

178 *"to reduce the risk":* Eli Lilly and Company, "Highlights of Prescribing Information," from label for Trulicity, revised September 2020, https://www.accessdata.fda.gov/drugsatfda_docs/label/2020/125469s036lbl.pdf.

a major selling point: Eli Lilly and Company, "Prescribing Information: Trulicity," Trulicity.com, https://www.trulicity.com/type-2-diabetes?utm_id=bi_cmp-291802755_adg-1264438970789604_ad-79027520074938_kwd-79027760368911:loc-190_dev-c_ext-&campaign=291802755&   adgroup=1264438970789604&ad=79027520074938&utm_keyword=kwd-79027760368911:loc-190&msclkid   =300eb73cba0a186aa62de862e6756f55&utm_source=bing&utm_medium=cpc&utm_campaign=US_DTC_Trulicity_Brand_CV_Exact&utm_term=trulicity%20cardiovascular&utm_content=Cardiovascular.

overall risk of death: Hertzel C. Gerstein et al., "Dulaglutide and Cardiovascular Outcomes in Type 2 Diabetes (REWIND): A Double-Blind, Randomised Placebo-Controlled Trial," *Lancet* 394 (2019): 121–30.

healthy lifestyle habits: Qiuhong Gong et al., "Morbidity and Mortality After Lifestyle Intervention for People with Impaired Glucose Tolerance: 30-Year Results of the Da Qing Diabetes Prevention Outcome Study," *Lancet: Diabetes and Endocrinology* 7, no. 6 (2019): 452–61.

179 *(like venture capital or):* Mehrsa Baradaran, "The Neoliberal Looting of America," *New York Times,* July 2, 2020, https://www.nytimes.com/2020/07/02/opinion/private-equity-inequality.html.

five hundred largest global: U.S. Government Accountability Office, "Drug Industry: Profits, Research and Development Spending, and Merger and Acquisition Deals," Report to Congressional Requesters, U.S. GAO, November 2017, https://www.gao.gov/assets/690/688472.pdf.

180 *biotech investors earned:* Sean Dickson and Jeromie Ballreich, "How Much Can Pharma Lose? A Comparison of Returns Between Pharmaceutical and Other Industries," West Health Policy Center, https://www.westhealth.org/wp-content/uploads/2019/11/WHPC_White-Paper_How-Much-Can-Pharma-Lose_FINAL-November-2019.pdf.

highest-revenue-generating: Paul B. Ginsburg and Steven M. Lieberman, "The Elijah E. Cummings Lower Drug Costs Now Act: How It Would Work, How It Would Affect Prices, and What the Challenges Are," Commonwealth Fund, April 9, 2020, https://www.commonwealthfund.org/publications/issue-briefs/2020/apr/lower-drug-costs-now-act-hr3-how-it-would-work.

a "nuclear winter": Jonathan Gardner, "House Passes Drug Pricing Bill That Pharma Warned Would Bring 'Nuclear Winter,'" *BiopharmaDive*, December 12, 2019, https://www.biopharma dive.com/news/house-ap proves-hr3-drug-pricing-bill-pharma/568966/.

$456 billion: Phillip L. Swagel, Congressional Budget Office, letter to Frank Pallone Jr., chairman, Committee on Energy and Commerce, U.S. House of Representatives, Re: "Budgetary Effects of H.R.3, the Elijah E. Cummings Lower Drug Costs Now Act," December 10, 2019, https://www.cbo.gov/system/files/2019-12/hr3_complete.pdf.

previously unavailable benefit: GAO, "Drug Industry."

181   *the next-highest industry*: Dickson and Ballreich, "How Much."

with the highest ROIC: Dickson and Ballreich, "How Much."

"socialist price controls": Chaarushena Deb and Gregory Curfman, "Relentless Prescription Drug Price Increases," *Journal of the American Medical Association* 323, no. 9 (2020): 826–28.

"America's Biopharmaceutical Companies": America's Biopharmaceutical Companies, "Member Companies," innovation.org, https://innova tion.org/about-us/members.

between 2016 and 2020: Staff Report Committee on Oversight and Reform, U.S. House of Representatives, "Drug Pricing Investigation Industry Spending on Buybacks, Dividends, and Executive Compensation," July 2021, oversight.house.gov.

182   *"Innovation might make"*: Jill Lepore, *These Truths: A History of the United States* (New York: W. W. Norton, 2018).

drugs that transformed HIV: Caroline A. Sabin, "Do People with HIV Infection Have a Normal Life Expectancy in the Era of Combination Antiretroviral Therapy?," *BMC Medicine* 11 (2013): 251, https://www.ncbi.nlm.nih.gov/pmc/articles/PMC4220799/pdf/1741-7015-11-251.pdf.

over $35,000 a year: GoodRx, "Atripla," https://www.goodrx.com/atri pla.

about $100 per year: Médecins Sans Frontières, *Untangling the Web of Antiretroviral Price Reductions*, 18th edition, July 2016, https://msfac cess.org/sites/default/files/HIV_report_Untangling-the-web-18thed_ENG_2016.pdf.

the first year of sales: Jerry Avorn, "The $2.6 Billion Pill — Methodologic and Policy Considerations," *New England Journal of Medicine* 372 (2015): 1877–79.

183 *$1,000 per pill:* Ketaki Gokhale, "The Same Pill That Costs $1,000 in the U.S. Sells for $4 in India," *Chicago Tribune,* January 4, 2016, https://www.chicagotribune.com/business/ct-drug-price-sofosbuvir-sovaldi-india-us-20160104-story.html.

49.7 percent profit: L. Marsa, "A 5-Point Plan to Lower Rx Prices," *AARP Bulletin* (May 2019), https://www.aarp.org/politics-society/advocacy/info-2019/5-point-prescription-drug-plan.html.

frequency of pulmonary infections: P. G. Middleton et al., "Elexacaftor-Tezacaftor-Ivacaftor for Cystic Fibrosis with a Single Phe508del Allele," *New England Journal of Medicine* 381 (2019): 1809–19, https://www.nejm.org/doi/pdf/10.1056/NEJMoa1908639?articleTools=true.

"Despite being transformative": "ICER Releases Evidence Report on Treatments for Cystic Fibrosis," *Institute for Clinical and Economic Review,* April 27, 2020, https://icer-review.org/announcements/cf_evidence_report_2020/.

reasonably cost-effective range: "Modulator Treatments for Cystic Fibrosis: Effectiveness and Value," Institute for Clinical and Economic Review, April 27, 2020, https://icer-review.org/wp-content/uploads/2019/09/ICER_CF_Evidence_Report_042720.pdf.

(more than $40,000): Mike Tigas et al., "Dollars for Docs: How Industry Dollars Reached Your Doctors," ProPublica, updated October 17, 2019, https://projects.propublica.org/docdollars/.

Canadians with cystic fibrosis: Anne L. Stephenson et al., "Survival Comparison of Patients with Cystic Fibrosis in Canada and the United States," *Annals of Internal Medicine* 166 (2017): 537–46.

184 *list of essential medicines:* "The Selection and Use of Essential Medicines," WHO Technical Report Series, unedited version (2019), 21, 168, 224, https://www.who.int/medicines/publications/essentialmedicines/UNEDITED_TRS_2019_EC22_Sept.pdf?ua=1.

"earned" in our country: Goldman and Lakdawalla, "The Global Burden of Medical Innovation."

185 *to believe that 80 percent:* Gordon B. Lindsay, Ray M. Merrill, and Riley J. Hedin, "The Contribution of Public Health and Improved Social Conditions to Increased Life Expectancy: An Analysis of Public Awareness," *Journal of Community Medicine and Health Education* 311 (2014), doi:10.4172/2161-0711.1000311.

186 *eleven other wealthy countries:* Kamal et al., "Health Spending and the Economy."

20 percent of our health: Magnan et al., "Achieving Accountability for Health and Health Care," *Minnesota Medicine,* 95 (2012): 37-9.

those who fund it: Moses et al., "The Anatomy of Medical Research."

population-based disease: David Satcher, "The Prevention Challenge and Opportunity," *Health Affairs* 25, no. 5 (2006): 1009–11.

187 *for all that's wrong:* Shrank, Rogstad, and Parekh, "Waste in the US Health Care System."

188 *unregulated and unjustified:* Emanuel, "The Real Cost."

and population health: Alan Weil, "The Social Determinants of Death," *Health Affairs Blog: Once in a Weil,* June 3, 2020, https://www.healthaffairs.org/do/10.1377/hblog20200603.831955/full/.

189 *"The dissolution of Haven":* Erin Brodwin and Casey Ross, "Inside the Collapse of a Disrupter: How Haven's High Hopes of Redefining Health Care Came to a Crashing Halt," *STAT+,* January 5, 2021, https://www.statnews.com/2021/01/05/haven-collapse-amazon-care-jpmorgan-berkshire/.

health-care businesses and hospitals: Anjalee Khemlani, "Why Berkshire Hathaway's Health Care Project Haven Failed," *yahoo! Finance,* April 24, 2021, https://finance.yahoo.com/news/why-warren-buffetts-health-care-project-haven-failed-111839197.html.

"From biotechnology": https://www.jpmorgan.com/commercial-banking/industries/life-sciences.

9. THE LIMITS OF OBAMACARE

193 *"Politically and emotionally":* Barack Obama, "A President Looks Back on His Toughest Fight," *New Yorker,* October 26, 2020.

"market society": Michael J. Sandel, "What Isn't for Sale?," *Atlantic,* April 2012, www.theatlantic.com/magazine/archive/2012/04/what-isnt-for-sale/308902/.

194 *"our broken health care":* "Baucus, Grassley Announce Health Reform Summit," U.S. Senate Committee on Finance, May 23, 2008, https://www.finance.senate.gov/release/ baucus-grassley-announce-health-reform-summit.

the American economy: Steven Brill, *America's Bitter Pill: Money, Politics, Backroom Deals, and the Fight to Fix Our Broken Healthcare System* (New York: Random House, 2015).

"Some people suggest": Brill, *America's Bitter Pill.*

195   *also supported the idea:* Brill, *America's Bitter Pill.*
      health care was expanded: Brill, *America's Bitter Pill.* At that time, forty-
      four million Americans were without health insurance (see https://www.
      kff.org/uninsured/ issue-brief/ key-facts-about-the-uninsured-popu
      lation/), in sharp contrast to the universal or near-universal insurance
      that had been achieved by all the other wealthy countries (see https://
      www.theatlantic.com/international/ archive/2012/06/heres-a-map-of-
      the-countries-that-provide-universal-health-care-americas-still-not-
      on-it/259153/).
      "You can either try to": Brill, *America's Bitter Pill.*
196   *"the insurance companies":* Brill, *America's Bitter Pill.*
      "If we don't address costs": Brill, *America's Bitter Pill.*
197   *"help people make informed":* Patient-Centered Outcomes Research In-
      stitute, https://www.pcori.org/about-us/our-vision-mission.
      "establish what type": P. J. Neumann and Milton C. Weinstein, "Legislat-
      ing Against Use of Cost-Effectiveness Information," *New England Journal
      of Medicine* 363 (2010): 1495–97.
      "activists on the left": Obama, "A President Looks Back."
198   *risk of collapsing:* Brill, *America's Bitter Pill.*
      Dow Jones Industrial Average: "Health Care Stocks: Performance Under
      Obamacare," healthcarereform.procon.org, https://healthcarereform.
      procon.org/health-care-stocks-performance-under-obamacare/.
      a low of 26.7 million: Jennifer Tolbert, Kendal Orgera, and Anthony
      Damico, "Key Facts About the Uninsured Population," Kaiser Fam-
      ily Foundation, November 6, 2020, https://www.kff.org/uninsured/
      issue-brief/key-facts-about-the-uninsured-population/.
      curtail health-care costs: Janet Weiner, Clifford Marks, and Mark Pauly,
      "Effects of the ACA on Health Care Cost Containment," Leonard Da-
      vis Institute of Health Economics, University of Pennsylvania, March 2,
      2017, https://ldi.upenn.edu/brief/effects-aca-health-care-cost-contain
      ment.
199   *"Economic elites":* Martin Gilens and Benjamin I. Page, "Testing Theo-
      ries of American Politics: Elites, Interest Groups, and Average Citizens,"
      Perspectives on Politics 12, no. 3 (2014): 564–81.
      cost of prescription drugs: Lunna Lopez et al., "KFF Health Tracking
      Poll — January 2020: Medicare-for-All, Public Option, Health Care Leg-
      islation and Court Actions," Kaiser Family Foundation, January 3, 2020,
      https://www.kff.org/health-reform/poll-finding/kff-health-tracking-
      poll-january-2020/.
      first eight months of 2020: Lev Facher, "Pharma Is Showering Congress

with Cash, Even as Drug Makers Race to Fight the Coronavirus," *STAT News,* August 10, 2020, https://www.statnews.com/feature/prescrip tion-politics/prescription-politics/.

refused to even bring it: Jonathan Cohn, "Democrats Did Something Big on Health Care and Nobody Is Paying Attention," *Huffington Post,* February 23, 2020, https://www.huffpost.com/entry/ prescription-drug-bill-house-democrats-hr3_n_5e4ddae5c5b 630e74c4ff4e3.

cost of their health care: Jeffrey M. Jones and R. J. Reinhart, "Americans Remain Dissatisfied with Healthcare Costs," Gallup, November 28, 2018, https://news.gallup.com/ poll/245054/americans-remain-dissatis fied-healthcare-costs.aspx.

adding a public option: Lopez et al., "Health Tracking Poll."

the 21st Century Cures Act: Sheila Kaplan, "Winners and Losers of the 21st Century Cures Act," *STAT News,* December 5, 2016, https:// www.statnews.com/2016/12/05/21st-century-cures-act-winners-losers/.

200   *(not for the drug companies):* "Q&A: 21st Century Cures Act May Favor 'Speed Over Science,'" *STAT News,* February 27, 2017, https://www.stat news.com/ pharmalot/2017/02/27/cures-act-science/.

"detecting and responding to": "Prevention and Public Health Fund," U.S. Department of Health and Human Services, last reviewed January 9, 2020, https://www.hhs.gov/open/prevention/index.html.

opposed the act: David Epstein, "When Evidence Says No, but Doctors Say Yes," ProPublica, February 22, 2017, https://www.propublica.org/ar ticle/when-evidence-says-no-but-doctors-say-yes.

"When American voters": Michael Hiltzik, "Column: The 21st Century Cures Act: A Huge Handout to the Drug Industry Disguised as a Pro-Research Bounty," *Los Angeles Times,* December 5, 2016, https://www.la times.com/business/hiltzik/la-fi-hiltzik-21st-century-20161205-story. html.

201   *more than two-thirds:* Facher, "Pharma Is Showering Congress."

12 percent in 2018: Sara R. Collins, Munira Z. Gunja, and Gabriella N. Aboulafia, "U.S. Health Insurance Coverage in 2020: A Looming Crisis in Affordability," Commonwealth Fund, August 19, 2020, https://www. commonwealthfund.org/publications/issue-briefs/2020/aug/loom ing-crisis-health-coverage-2020-biennial.

actually decreased: "U.S. Life Expectancy 1950–2021," Macrotrends.

expanding coverage and containing costs: Brill, *America's Bitter Pill.*

10. THE KEY TO MEANINGFUL REFORM

202 *"Corporations have a valuable"*: Binyamin Appelbaum, "50 Years of Blaming Milton Friedman: Here's Another Idea," *New York Times,* September 18, 2020, https://www.nytimes.com/2020/09/18/opinion/milton-friedman-essay.html.

204 *more than 8.5 percent:* Joe Biden, "Health Care," https://joebiden.com/healthcare/#.

same vested interests: Peter Sullivan, "Health Care Industry Launches New Ads Against Public Option for Convention," *Hill,* August 14, 2020, https://thehill.com/policy/healthcare/512003-health-care-industry-launches-new-ads-against-public-option-for-convention.

candidate Biden's proposal: Andrew Perez and David Sirota, "Democrats Seem All Too Willing to Surrender on Health Care Reform," *Jacobin,* August 2020, https://www.jacobinmag.com/2020/08/ health-care-public-option-joe-biden-aca.

"Economists agree that": Partnership for America's Health Care Future, "Priorities," https://americashealthcarefuture.org/priorities/.

205 *more than a low income:* Dylan Scott, "Joe Biden's Health Care Plan, Explained in 800 Words," *Vox,* November 6, 2020, https://www.vox.com/21540041/election-2020-joe-biden-health-care.

challenge the vested interests: Sarah Kliff, "Private Insurance Wins in Democrats' First Try at Expanding Health Coverage," *New York Times,* March 5, 2021.

"What we won't accept": Nicholas Florko, "PhRMA's New Message to Washington: Don't Take Us for Granted," *STAT+,* April 13, 2021.

206 *their PR buddies:* Dylan Scott, "Why Democrats' Ambitions for Health Care Are Shrinking Rapidly," *Vox,* May 7, 2021.

current disjointed hodgepodge: Emily Gee and Topher Spiro, "Excess Administrative Costs Burden the U.S. Health Care System," Center for American Progress, April 8, 2019, https://www.americanprogress.org/issues/healthcare/reports/2019/04/08/468302/excess-administrative-costs-burden-u-s-health-care-system/.

Shumlin withdrew the plan: Joe VerValin, "The Rise and Fall of Vermont's Single Payer Plan," *Cornell Policy Review,* July 13, 2017, http://www.cornellpolicyreview.com/rise-fall-vermonts-single-payer-plan/.

207 *sank the plan:* John E. McDonough, "The Demise of Vermont's Single-Payer Plan," *New England Journal of Medicine* 372 (2015): 1584–85.

208 *"health and well-being":* Moses et al., "The Anatomy of Medical Research."

209 *standards of research ethics:* Bošnjak Snežana, "The Declaration of Helsinki: The Cornerstone of Research Ethics," *Archive of Oncology* 9, no. 3 (2001): 179–84, http://scindeks.ceon.rs/article.aspx?artid=0354-73100103179B&lang=en.

"the benefits, risks, burdens": "World Medical Association Declaration of Helsinki: Ethical Principles for Medical Research Involving Human Subjects," *Journal of the American Medical Association* 310, no. 20 (2013): 2191–94, https://jamanetwork.com/journals/jama/fullarticle/1760318.

210 *(Nazi concentration camps):* John R. Williams, "The Declaration of Helsinki and Public Health," SciELO Public Health, Bulletin of the World Health Organization, June 11, 2008, https://www.scielosp.org/article/bwho/2008.v86n8/650-652/.

211 *a single "master protocol":* Carl Zimmer, "First, a Vaccine Approval. Then 'Chaos and Confusion,'" *New York Times,* October 12, 2020, https://www.nytimes.com/2020/10/12/health/covid-vaccines.html.

212 *within one year:* "All Trials Registered / All Results Reported," AllTrials.net, https://www.alltrials.net/find-out-more/all-trials/#_ftn8.

213 *"is poor, and not improving":* Nicholas J. DeVito, Seb Bacon, and Ben Goldacre, "Compliance with Legal Requirement to Report Clinical Trial Results on Clinicaltrials.Gov: A Cohort Study," *Lancet* 395 (2020): 361–69, https://www.thelancet.com/journals/lancet/article/PIIS0140-6736(19)33220-9/fulltext.

"to support regulators": "Joint Statement on Transparency and Data Integrity, International Coalition of Medicines Regulatory Authorities (IC-MRA) and WHO," *World Health Organization,* May 7, 2021.

"The FDA understands": Ed Silverman, "WHO, International Regulatory Group Urge Pharma to Publish Clinical Study Reports," *STAT+,* May 7, 2021.

214 *New Zealand has the lowest:* Steven G. Morgan, Christine Leopold, and Anita K. Wagner, "Drivers of Expenditure on Primary Care Prescription Drugs in 10 High-Income Countries with Universal Health Coverage," *Canadian Medical Association Journal,* June 12, 2017, 189: E794-9. (U.S. added).

215 *"Most people taking":* Trulicity commercial, 2021, https://www.bing.com/videos/search?q=trulicity+tv+ads+2021&view=detail&mid=4A8851E3FD4ACDEB53AF4A8851E3FD4ACDEB53AF&FORM=VIRE

between 2013 and 2020: For example, Beth Snyder Bulik, "The Top 10 Ad Spenders in Big Pharma for 2020," *FiercePharma,* April 19, 2021.

almost $78,000 in 2021: "Drug Pricing Investigation AbbVie — Humira and Imbruvica," *Staff Report Committee on Oversight and Reform, U.S. House of Representatives,* May 2021, oversight.house.gov.

216   *"to preserve law":* Friedman, *Capitalism and Freedom,* 2.

but not in fact: Jed S. Rakoff, "Getting Away with Murder," *New York Review of Books,* December 3, 2020.

218   *"based on additional":* Aaron Kesselheim, "High Drug Prices in the US: What We Can Learn from Other Countries (and Some U.S. States)," Testimony to the United States Senate Committee on Health, Education, Labor, and Pensions, March 23, 2021, Washington, DC.

just four areas: Emanuel, "The Real Cost."

219   *"competent, unbiased information":* Celia Wexler, "Bring Back the Office of Technology Assessment," *New York Times,* May 28, 2015, https://www.nytimes.com/roomfordebate/2015/05/28/scientists-curbing-the-ethical-use-of-science/bring-back-the-office-of-technology-assessment.

a health program: David Banta and Clyde J. Behney, "Office of Technology Assessment Health Program," *International Journal of Technology Assessment in Health Care* 25 (2009): 28–32.

health technology assessment: John C. O'Donnell et al., "Health Technology Assessment: Lessons Learned from Around the World — an Overview," *ScienceDirect* 12, suppl 2 (2009): S1–S5, https://www.sciencedirect.com/science/article/pii/S109830151060054X.

"Contract with America": David Malakoff, "House Democrats Move to Resurrect Congress's Science Advisory Office," *Science,* April 30, 2019, https://www.sciencemag.org/news/2019/04/house-democrats-move-resurrect-congress-s-science-advisory-office.

no government agency: Goldman, Nussbaum, and Linthicum, "Rapid Biomedical Innovation."

prohibited from considering: Victor R. Fuchs, "Health Care Policy After the COVID-19 Pandemic," *Journal of the American Medical Association* 324, no. 3 (2020): 233–34, https://jamanetwork.com/journals/jama/fullarticle/2767352.

"clinical excellence in": "The National Institute for Clinical Excellence (Establishment and Constitution) Order 1999," National Health Service, England and Wales, Statutory Instruments, 1999, https://www.legislation.gov.uk/uksi/1999/220/pdfs/uksi_19990220_en.pdf.

remains at arm's length: Corinna Sorenson et al., "National Institute for Health and Clinical Excellence (NICE): How Does It Work and What Are the Implications for the U.S.?," *National Pharmaceutical Council,* May 2008.

220 *underlying clinical trial data:* "Appendix A: Summary of Evidence from Surveillance," NICE, September 27, 2016, 52–64, https://www.nice.org .uk/guidance/cg181/evidence/appendix-a-evidence-summaries-pdf -4724759774.

their conflicts of interest: Kate L. Mandeville et al., "Financial Interests of Patient Organisations Contributing to Technology Assessment at England's National Institute for Health and Care Excellence: Policy Review," *British Medical Journal* 364 (2019): k5300.

price remains confidential: Sarah Boseley, "NHS England Agrees Price for 'Unaffordable' Cystic Fibrosis Drug," *Guardian,* October 24, 2019, https://www.theguardian.com/society/2019/oct/24/ nhs-england-ver tex-agrees-price-for-orkambi-unaffordable-cystic-fibrosis-drug.

experience of UK citizens: NICE, "Data Collection Agreement," National Institute for Health and Care Excellence, https://www.nice.org.uk/ about/what-we-do/our-programmes/nice-guidance/nice-technology -appraisal-guidance/data-collection-agreement.

221 *2.7 years longer:* Roosa Tikkanen, "Multinational Comparisons of Health Systems Data, 2019," Commonwealth Fund, 2019, https://www .commonwealthfund.org/publications/other-publication/2020/jan/ multinational-comparisons-health-systems-data-2019.

is ranked last: Schneider et al., "Mirror, Mirror 2017."

nonmedical determinants: Elizabeth H. Bradley et al., "Health and Social Services Expenditures: Associations with Health Outcomes," *BMJ Quality and Safety* 20 (2011): 826e831.

222 *"We know . . . that Congress":* Brill, *America's Bitter Pill.*

11. REFORM FROM THE BOTTOM UP

223 *"The hard fact is":* David Blumenthal and Margaret Hamburg, "US Health and Health Care Are a Mess: Now What?," *Lancet,* February 11, 2021, https://doi.org/10.1016/S0140-6736(21)00318-4/.

intervention in our economy: Reich, *Saving Capitalism.*

"to do what was right": "Public Trust in Government: 1958–2019," Pew Research Center, April 11, 2019.

224 *(near an all-time low):* "Public Trust in Government."

declined in absolute terms: Steven H. Woolf, Ryan K. Masters, and Laudan Y. Aron, "Effect of the Covid-19 Pandemic in 2020 on Life Expectancy Across Populations in the USA and Other High Income Countries: Simulations of Provisional Mortality Data," *BMJ*, 2021, 373: n1343.

225 *has increased thirtyfold:* CMS.gov, "National Health Expenditures Table 16: Retail Prescription Drugs Expenditures; Levels, Percent Change, and Percent Distribution, by Source of Funds: Selected Calendar Years, 1970-2019," Historical, CMS, December 16, 2020.

"more about profits": Lisa Ellis, "Snapshot of the American Pharmaceutical Industry," Harvard T. H. Chan School of Public Health, July 14, 2016, https://www.hsph.harvard.edu/ecpe/snapshot-of-the-american-pharmaceutical-industry/.

most poorly regarded: Justin McCarthy, "Big Pharma Sinks to the Bottom of U.S. Industry Rankings," Gallup, September 3, 2019, https://news.gallup.com/poll/266060/big-pharma-sinks-bottom-industry-rankings.aspx.

AFTERWORD

231 *it had been in 1998:* Woolf et al., "Effect of the Covid-19 Pandemic," 2021.

40 percent more likely: Nambi Ndugga, Olivia Pham, Latoya Hill et al., "Latest Data on COVID-19 Vaccinations by Race/Ethnicity," Kaiser Family Foundation, June 30, 2021.

232 *statistically very unlikely:* David S. Knopman, David T. Jones, Michael D. Greicius, "Failure to Demonstrate Efficacy of Aducanumab: An Analysis of the EMERGE and ENGAGE Trials as Reported by Biogen, December 2019," *Alzheimer's Dement* 17 (2021): 696–701.

to be clinically *significant:* G. Caleb Alexander, Scott Emerson, and Aaron S. Kesselheim, "Evaluation of Aducanumab for Alzheimer Disease: Scientific Evidence and Regulatory Review Involving Efficacy, Safety, and Futility," *Journal of the American Medical Association* 325 (2021): 1717–18.

Biogen's own safety data: Samantha Budd Haeberlein, Christian von Hehn, Ying Tian et al., "EMERGE and ENGAGE Topline Results: Two Phase 3 Studies to Evaluate Aducanumab in Patients with Early Alzheimer's Disease," Biogen, Cambridge, MA, https://investors.biogen.com/static-files/8e58afa4-ba37-4250-9a78-2ecfb63b1dcb#_blank.

"In summary, the totality": Tristan Massie, FDA Statistical Review and Evaluation, Aducanumab, July 7, 2020. https://www.accessdata.fda.gov/drugsatfda_docs/nda/2021/761178Orig1s000StatR_Redacted.pdf.

233 *"primary evidence of":* Final Summary Minutes of the Peripheral and Central Nervous System Drugs Advisory Committee Meeting, Food and

Drug Administration Center for Drug Evaluation and Research, November 6, 2020.

"fairly confident": Laurie McGinley, "FDA Releases Fresh Details on Internal Debate over Controversial Alzheimer's Drug," *Washington Post,* June 22, 2021.

or had shown toxicity: F. Panza, M. Lozupone, G. Logroscino et al., "A Critical Appraisal of Amyloid-β-Targeting Therapies for Alzheimer Disease," *Nature Reviews Neurology* 15 (2019): 73–88.

"no sufficiently reliable": Early Alzheimer's Disease: Developing Drugs for Treatment, Guidance for Industry, FDA Center for Drug Evaluation and Research, February 2018, https://www.fda.gov/media/110903/download.

"reduce the emotional": Beth Wang, "Cavazonni: FDA May Loosen Advisory Committee Conflict of Interest Rules," *Inside Washington,* June 15, 2021.

234 *(adding $57 billion):* Juliette Cubanski and Tricia Neuman, "FDA's Approval of Biogen's New Alzheimer's Drug Has Huge Cost Implications for Medicare and Beneficiaries," Kaiser Family Foundation, June 10, 2021.

"this drug will be effective": STAT–Harris Poll, Fielding Period: June 11–13, 2021, http://freepdfhosting.com/281957e09e.pdf.

"We have met the enemy": Walt Kelly, "Pogo," April 22, 1971.

INDEX